The Principles of Animal Breeding
The Physiological Laws Involved in the Reproduction and Improvement of Domestic Animals

by S.L. Goodale

with an introduction by Jackson Chambers

Self Reliance Books

Get more historic titles on animal and stock breeding, gardening and old fashioned skills by visiting us at:

http://selfreliancebooks.blogspot.com/

Introduction

I am pleased to present yet another title on the Principles of Animal Breeding.

This volume is entitled "Principles of Breeding" and was published in 1861.

The work is in the Public Domain and is re-printed here in accordance with Federal Laws.

As with all reprinted books of this age that are intended to perfectly reproduce the original edition, considerable pains and effort had to be undertaken to correct fading and sometimes outright damage to existing proofs of this title. At times, this task is quite monumental, requiring an almost total "rebuilding" of some pages from digital proofs of multiple copies. Despite this, imperfections still sometimes exist in the final proof and may detract from the visual appearance of the text.

I hope you enjoy reading this book as much as I enjoyed making it available to readers again.

Jackson Chambers

PREFACE.

THE writer has had frequent occasion to notice the want of some handy book embodying the principles necessary to be understood in order to secure improvement in Domestic Animals.

It has been his aim to supply this want.

In doing so he has availed himself freely of the knowledge supplied by others, the aim being to furnish a useful, rather than an original book.

If it serve in any measure to supply the need, and to awaken greater interest upon a matter of vital importance to the agricultural interests of the country, the writer's purpose will be accomplished.

CONTENTS.

THE
PRINCIPLES OF BREEDING.

————•◆•————

CHAPTER I.

INTRODUCTORY.

The object of the husbandman, like that of men engaged in other avocations, is *profit*; and like other men the farmer may expect success proportionate to the skill, care, judgment and perseverance with which his operations are conducted.

The better policy of farmers generally, is to make stock husbandry in some one or more of its departments a leading aim—that is to say, while they shape their operations according to the circumstances in which they are situated, these should steadily embrace the conversion of a large proportion of the crops grown into animal products,—and this because, by so doing, they may not only secure a present livelihood, but best maintain and increase the fertility of their lands.

The object of the stock grower is to obtain the most valuable returns from his vegetable products. He

2

needs, as Bakewell happily expressed it, "the best *machine* for converting herbage and other animal food into money."

He will therefore do well to seek such animals as are most perfect of their kind—such as will pay best for the expense of procuring the machinery, for the care and attention bestowed, and for the consumption of raw material. The returns come in various forms. They may or may not be connected with the ultimate value of the animal. In the beef ox and the mutton sheep, they are so connected to a large extent; in the dairy cow and the fine wooled sheep, this is quite a secondary consideration;—in the horse, valued as he is for beauty, speed and draught, it is not thought of at all.

Not only is there a wide range of field for operations, from which the stock grower may select his own path of procedure, but there is a demand that his attention be directed *with a definite aim,* and *towards an end clearly apprehended.* The first question to be answered, is, what do we want? and the next, how shall we get it?

What we want, depends wholly upon our situation and surroundings, and each must answer it for himself. In England the problem to be solved by the breeder of neat cattle and sheep is how " to produce an animal or a living machine which with a certain quantity and quality of food, and under certain given circumstances, shall yield in the shortest time the largest quantity and best quality of beef, mutton or milk, with the largest profit to the producer and at least cost to the consumer." But this is not precisely the problem for American farmers to solve, because our circumstances are different. Few, if any, here grow oxen for beef alone, but for labor and beef, so that earliest possible maturity may be omitted and a year or more of labor profitably intervene before conversion to beef. Many cultivators of sheep, too, are so situated as to prefer fine wool, which is incompatible with the largest quantity and best quality of meat. Others differently situated in regard to a meat market would do well to follow the English practice and aim at the most profitable production of mutton. A great many farmers, not only of those in the vicinity of large towns, but of those at some distance, might, beyond doubt, cultivate dairy qualities in cows, to great advantage, and this too, even, if necessary, at the sacrifice, to considerable extent, of beef making qualities. As a general thing dairy qualities have been sadly neglected in years past.

Whatever may be the object in view, it should be clearly apprehended, and striven for with persistent and well directed efforts. To buy or breed common animals of mixed qualities and use them for any and for all purposes is too much like a manufacturer of cloth procuring some carding, spinning and weaving machinery, adapted to no particular purpose but which can somehow be used for any, and attempting to make fabrics of cotton, of wool, and of linen with it. I do not say that cloth would not be produced, but he would assuredly be slow in getting rich by it.

The stock grower needs not only to have a clear and definite aim in view, but also to understand the means by which it may best be accomplished. Among these means a knowledge of the principles of breeding holds a prominent place, and this is not of very easy acquisition by the mass of farmers. The experience of any one man would go but a little way towards acquiring it, and there has not been much published on the subject in any form within the reach of most. I have been able to find nothing like an extended systematic treatise on the subject, either among our own or the foreign agricultural literature which has come within my notice. Indeed, from the scantiness of what appears to have been written, coupled with the fact that much knowledge must exist somewhere, one is tempted to believe that not all which might have done so, has yet found its way to printers' ink. That a great deal has been acquired, we know, as we know a tree—by its fruits. That immense achievements have been accomplished is beyond doubt.

The improvement of the domestic animals of a country so as greatly to enhance their individual and aggregate value, and to render the rearing of them more profitable to all concerned, is surely one of the achievements of advanced civilization and enlightenment, and is as much a triumph of science and skill as the construction of a railroad, a steamship, an electric telegraph, or any work of architecture. If any doubt this, let them ponder the history of those breeds of animals which have made England the stock nursery of the world, the perfection of which enables her to export thousands of animals at prices almost fabulously beyond their value for any purpose but to propagate their kind; let them note the patient industry, the genius and application which have been put forth to bring them to the condition they have attained, and their doubts must cease.

Robert Bakewell of Dishley, was one of the first of these improvers. Let us stop for a moment's glance at him. Born in 1725, on the farm where his father and grandfather had been tenants, he began at the age of thirty to carry out the plans for the improvement of domestic animals upon which he had resolved as the result of long and patient study and reflection. He was a man of genius, energy and perseverance. With sagacity to conceive and fortitude to perfect his designs, he laid his plans and struggled against many disappointments, amid the ridicule and predictions of failure freely bestowed by his neighbors,—often against serious pecuniary embarrassments ; and at last was crowned by a wonderful degree of success. When he commenced letting his rams, (a system first introduced by him and adhered to during his life, in place of selling,) they brought him 17s. 6d. each, for the season. This was ten years after he commenced his improvements. Soon the price came to a guinea, then to two or three guineas—rapidly increasing with the reputation of his stock, until in 1784, they brought him 100 guineas each ! Five years later his lettings for one season amounted to $30,000 !

With all his skill and success he seemed afraid lest others might profit by the knowledge he had so laboriously acquired. He put no pen to paper and at death left not even the slightest memorandum throwing light upon his operations, and it is chiefly through his cotemporaries, who gathered somewhat from verbal communications, that we know anything regarding them. From these we learn that he formed an ideal standard in his own mind and then endeavored, first by a wide selection and a judicious and discriminating coupling, to obtain the type desired, and then by close breeding, connected with rigorous weeding out, to perpetuate and fix it.

After him came a host of others, not all of whom concealed their light beneath a bushel. By long continued and extensive observation, resulting in the collection of numerous facts, and by the collation of these facts of nature, by scientific research and practical experiments, certain physiological laws have been discovered, and principles of breeding have been deduced and established. It is true that some of these laws are as yet hidden from us, and much regarding them is but imperfectly understood. What we do not know is a deal more than what we do know, but to ignore so much as has been discovered, and is well established, and can be learned by any who care to do so, and to go on regardless of it, would indicate a degree of wisdom in the breeder on a par with that of a builder who should fasten together wood and iron just as the pieces happened to come to his hand, regardless of the laws of architecture, and expect a convenient house or a fast sailing ship to be the result of his labors.

Is not the usual course of procedure among many farmers too nearly parallel to the case supposed? Let the ill-favored, chance-bred, mongrel beasts in their barn yards testify. The truth is, and it is of no use to deny or disguise the fact, the *improvement* of domestic animals is one of the most important and to a large extent, one of the most neglected branches of rural economy. The fault is not that farmers do not keep stock enough, much oftener they keep more than they can feed to the most profitable point, and when a short crop of hay comes, there is serious difficulty in supporting them, or in selling them at a paying price; but the great majority neither bestow proper care upon the selection of animals for breeding, nor do they appreciate the dollars and cents difference between such as are profitable and such as are profitless. How many will hesitate or refuse to pay a dollar for the services of a good bull when some sort of a calf can be begotten for a " quarter ?" and this too when one by the good male would be worth a dollar more for veal and ten or twenty dollars more when grown to a cow or an ox ? How few will hesitate or refuse to allow to a butcher the cull of his calves and lambs for a few extra shillings, and this when the butcher's difference in shillings would soon, were the best kept and the worst sold, grow into as many dollars and more ? How many there are who esteem size to be of more consequence than symmetry, or adaptation to the use for which they are kept? How many ever sit down to calculate the difference in money value between an animal which barely pays for keeping, or perhaps not that, and one which pays a profit?

Let us reckon a little. Suppose a man wishes to buy a cow. Two are offered him, both four years old, and which might probably be serviceable for ten years to come. With the same food and attendance the first will yield for ten months in the year, an average of five quarts per day,—and the other for the same term will yield seven quarts and of equal quality. What is the comparative value of each? The difference in yield is six hundred quarts per annum. For the purpose of this calculation we will suppose it worth three cents per quart—amounting to eighteen dollars. Is not the second cow, while she holds out to give it, as good as the first, and three hundred dollars at interest besides? If the first just pays for her food and attendance, the second, yielding two-fifths more, pays *forty per cent. profit* annually; and yet how many farmers having two such cows for sale would make more than ten, or twenty, or at most, thirty dollars difference in the price? The profit from one is eighteen dollars a year—in ten years one hundred and eighty dollars, besides the annual accumulations of interest—the profit of the other is—nothing. If the seller has need to keep one, would he not be wiser to give away the first, than to part with the second for a hundred dollars?

Suppose again, that an acre of grass or a ton of hay costs five dollars, and that for its consumption by a given set of animals, the farmer gets a return of five dollars worth of labor, or meat, or wool, or milk. He is selling his crop at cost, and makes no profit. Suppose by employing other animals, better horses, better cows, oxen and sheep, he can get ten dollars per ton in returns. How much are the latter worth more than the former? Have they not doubled the value of the crops, and increased the profit of farming from nothing to a hundred per cent? Except that the manure is not doubled, and the animals would some day need to be replaced, could he not as well afford to give the price of his farm for one set as to accept the other as a gift?

Among many, who are in fact ignorant of what goes to constitute merit in a breeding animal, there is an inclination to treat as imaginary and unreal the higher values placed upon well-bred animals over those of mixed origin, unless they are larger and handsomer in proportion to the price demanded. The sums paid for qualities which are not at once apparent to the eye are stigmatized as *fancy prices*. It is not denied that fancy prices are sometimes, perhaps often paid, for there are probably few who are not willing occasionally to pay dearly for what merely pleases them, aside from any other merit commensurate to the price.

But, on the other hand, it is fully as true that great

intrinsic value for breeding purposes may exist in an animal and yet make very little show. Such an one may not even look so well to a casual observer, as a grade, or cross-bred animal, which although valuable as an individual, is not, for breeding purposes, worth a tenth part as much.

Let us suppose two farmers to need a bull; they go to seek and two are offered, both two years old, of similar color, form and general appearance. One is offered for twenty dollars—for the other a hundred is demanded. Satisfactory evidence is offered that the latter is no better than any or all of its ancestors for many generations back on both sides, or than its kindred—that it is of a pure and distinct breed, that it possesses certain well known hereditary qualities, that it is suited for a definite purpose, it may be a Shorthorn, noted for large size and early maturity, it may be a Devon, of fine color and symmetry, active and hardy, it may be an Ayrshire, noted for dairy qualities, or of some other definite breed, whose uses, excellencies and deficiencies are all well known.

The other is of no breed whatever, perhaps it is called a grade or a cross. The man who bred it had rather confused ideas, so far as he had any, about breeding, and thought to combine all sorts of good qualities in one animal, and so he worked in a little grade Durham, or Hereford to get size, and a little Ayrshire for milk, and a little Devon for color, and so on, using perhaps dams sired by a bull in the neighborhood which had also got some "Whitten"* or "Peter Waldo" calves, (though none of these showed it,) at any rate he wanted some of the "native" element in his stock, because it was tough, and some folks thought natives were the best after all. Among its ancestors and kindred were some good and some not good, some large and some small, some well favored and fat, some ill favored and lean, some profitable and some profitless. The animal now offered is a great deal better than the average of them. It looks for aught they can see, about as well as the one for which five times his price is asked. Perhaps he served forty cows last year and brought his owner as many quarters, while the other only served five and brought an income of but five dollars. The question arises, which is the better bargain? After pondering the matter, one buys the low-priced and the other the high-priced one, both being well satisfied in their own minds.

What did results show? The low-priced one served that season perhaps a hundred cows; more than ought to have done so, came a second time;—having been overtasked as a yearling, he lacked somewhat of vigor.

---

*Local names for *lyery*, or black fleshed cattle.

The calves came *of all sorts*, some good, some poor, a few like the sire, more like the dams—all mongrels and showing mongrel origin more than he did. There seemed in many of them a tendency to combine the defects of the grades from which he sprung rather than their good points. In some, the quietness of the Short-horn degenerated into stupidity, and in others the activity of the Devon into nervous viciousness. Take them together they perhaps paid for rearing, or nearly so. After using him another year, he was killed, hav-ing been used long enough.

The other, we will say, served that same season a reasonable number, perhaps four to six in a week, or one every day, not more. Few came a second time and those for no fault of his. The calves bear a striking resemblance to the sire. Some from the better cows look even better in some points, than himself and few much worse. There is a remarkable uniformity among them; as they grow up they thrive better than those by the low priced one. They prove better adapted to the use intended. On the whole they are quite satis-factory and each pays annually in its growth, labor or milk a profit over the cost of food and attendance of five or ten dollars or more. If worked enough to fur-nish the exercise needful to insure vigorous health, he may be as serviceable and as manageable at eight or

3

ten years old, as at two; meantime he has got, perhaps, five hundred calves, which in due time become worth ten or twenty dollars each more than those from the other. Which now seems the wiser purchase? Was the higher estimate placed on the well bred animal based upon fancy or upon intrinsic value?

The conviction that a better knowledge of the principles of breeding would render our system of agriculture more profitable, and the hope of contributing somewhat to this end, have induced the attempt to set forth some of the physiological principles involved in the reproduction of domestic animals, or in other words, the laws which govern hereditary transmission.

CHAPTER II.

The Law of Similarity.

The first and most important of the laws to be considered in this connection is that of SIMILARITY. It is by virtue of this law that the peculiar characters, qualities and properties of the parents, whether external or internal, good or bad, healthy or diseased, are transmitted to their offspring. This is one of the plainest and most certain of the laws of nature. Children resemble their parents, and they do so because these are hereditary. The law is constant. Within certain limits progeny always and every where resemble their parents. If this were not so, there would be no constancy of species, and a horse might beget a calf or a sow have a litter of puppies, which is never the case,— for in all time we find repeated in the offspring the structure, the instincts and all the general characteristics of the parents, and never those of another species. Such is the law of nature and hence the axiom that "like produces like." But while experience teaches the constancy of hereditary transmission, it teaches just as plainly that the constancy is not absolute and

perfect, and this introduces us to another law, viz:
that of variation, which will be considered by and by;
our present concern is to ascertain what we can of the
law of similarity.

The lesson which this law teaches might be stated in
five words, to wit: *Breed only from the best*—but the
teaching may be more impressive, and will more likely
be heeded, if we understand the extent and scope of the
law.

Facts in abundance show the hereditary tendency of
physical, mental and moral qualities in men, and very
few would hesitate to admit that the external form and
general characteristics of parents descend to children
in both the human and brute races; but not all are
aware that this law reaches to such minute particulars
as facts show to be the case.

We see hereditary transmission of a peculiar type
upon an extensive scale, in some of the distinct races,
the Jews, and the Gypsies, for example. Although
exposed for centuries to the modifying influences of
diverse climates, to association with peoples of widely
differing customs and habits, they never merge their
peculiarities in those of any people with whom they
dwell, but continue distinct. They retain the same
features, the same figures, the same manners, customs
and habits. The Jew in Poland, in Austria, in London,

or in New York, is the same ; and the money-changers
of the Temple at Jerusalem in the time of our Lord
may be seen to-day on change in any of the larger
marts of trade. How is this? Just because the Jew
is a "thorough-bred." There is with him no intermarriage with the Gentile—no crossing, no mingling of his
organization with that of another. When this ensues
"permanence of race" will cease and give place to
variations of any or of all sorts.

Some families are remarkable during long periods
for tall and handsome figures and striking regularity of
features, while in others a less perfect form, or some
peculiar deformity reappears with equal constancy. A
family in Yorkshire is known for several generations to
have been furnished with six fingers and toes. A family
possessing the same peculiarity resides in the valley of
the Kennebec, and the same has reappeared in one or
more other families connected with it by marriage.

The thick upper lip of the imperial house of Austria,
introduced by the marriage of the Emperor Maximillian
with Mary of Burgundy, has been a marked feature in
that family for hundreds of years, and is visible in their
descendants to this day. Equally noticeable is the
"Bourbon nose" in the former reigning family of
France. All the Barons de Vessins had a peculiar
mark between their shoulders, and it is said that by

3*

means of it a posthumous son of a late Baron de Vessins was discovered in a London shoemaker's apprentice. Haller cites the case of a family where an external tumor was transmitted from father to son which always swelled when the atmosphere was moist.

A remarkable example of a singular organic peculiarity and of its transmission to descendants, is furnished in the case of the English family of "Porcupine men," so called from having all the body except the head and face, and the soles and palms, covered with hard dark-colored excrescences of a horny nature. The first of these was Edward Lambert, born in Suffolk in 1718, and exhibited before the Royal Society when fourteen years of age. The other children of his parents were naturally formed; and Edward, aside from this peculiarity, was good looking and enjoyed good health. He afterward had six children, all of whom inherited the same formation, as did also several grand-children.

Numerous instances are on record tending to show that even accidents do sometimes, although not usually, become hereditary. Blumenbach mentions the case of a man whose little finger was crushed and twisted by an accident to his right hand. His sons inherited right hands with the little finger distorted. A bitch had her hinder parts paralyzed for some days by a blow. Six of her seven pups were deformed, or so weak in

their hinder parts that they were drowned as useless.
A pregnant cat got her tail injured ; in each of her five
kittens the tail was distorted, and had an enlargement
or knob near the end of each.   Horses marked during
successive generations with red-hot irons in the same
place, transmit visible traces of such marks to their
colts.

Very curious are the facts which go to show that
acquired habits sometimes become hereditary.   Pritch-
ard, in his "Natural History of Man," says that the
horses bred on the table lands of the Cordilleras "are
carefully taught a peculiar pace which is a sort of
running amble ;" that after a few generations this
pace becomes a natural one ; young untrained horses
adopting it without compulsion.   But a still more
curious fact is, that if these domesticated stallions breed
with mares of the wild herd, which abound in the sur-
rounding plains, they "become the sires of a race in
which the ambling pace is natural and requires no
teaching."

Mr. T. A. Knight, in a paper read before the Royal
Society, says, "the hereditary propensities of the offspring of Norwegian ponies, whether full or half-bred,
are very singular.   Their ancestors have been in the
habit of *obeying the voice* of their riders and *not the bri-
dle* ; and horse-breakers complain that it is impossible

to produce this last habit in the young colts. They are, however, exceedingly docile and obedient when they understand the commands of their masters."

A late writer in one of the foreign journals, says that he had a "pup taken from its mother at six weeks old, who although never taught to 'beg' (an accomplishment his mother had been taught) spontaneously took to begging for every thing he wanted when about seven or eight months old ; he would beg for food, beg to be let out of the room, and one day was found opposite a rabbit hutch apparently begging the rabbits to come and play."

If even in such minute particulars as these, hereditary transmission may be distinctly seen, it becomes the breeder to look closely to the "like" which he wishes to see reproduced. Judicious selection is indispensable to success in breeding, and this should have regard to *every particular*—general appearance, length of limb, shape of carcass, development of chest; if in cattle, the size, shape and position of udder, thickness of skin, "touch," length and texture of hair, docility, &c., &c. ; if in horses, their adaptation to any special excellence depending on form, or temperament, or nervous energy.

Not only should care be taken to avoid *structural defects*, but especially to secure freedom from *hereditary diseases,* as both defects and diseases appear to be more easily transmissible than desirable qualities. There is often no obvious peculiarity of structure, or appearance, indicating the possession of diseases or defects which are transmissible, and so, special care and continued acquaintance are necessary in order to be assured of their absence in breeding animals; but such a tendency although invisible or inappreciable to cursory observation, must still, judging from its effects, have as real and certain an existence, as any peculiarity of form or color.

Every one who believes that a disease may be hereditary at all, must admit that certain individuals possess certain tendencies which render them especially liable to certain diseases, as consumption or scrofula; yet it is not easy to say precisely in what this predisposition consists. It seems probable, however, that it may be due either to some want of harmony between different organs, some faulty formation or combination of parts, or to some peculiar physical or chemical condition of the blood or tissues; and that this altered state, constituting the inherent congenital tendency to the disease, is duly transmitted from parent to offspring like any other quality more readily apparent to observation.

Hereditary diseases exhibit certain eminently characteristic phenomena, which a late writer* enumerates as follows :

1. "They are transmitted by the male as well as by the female parent, and are doubly severe in the offspring of parents both of which are affected by them.

2. They develop themselves not only in the immediate progeny of one affected by them, but also in many subsequent generations.

3. They do not, however, always appear in each generation in the same form ; one disease is sometimes substituted for another, analogous to it, and this again after some generations becomes changed into that to which the breed was originally liable—as phthisis (consumption) and dysentery. Thus, a stock of cattle previously subject to phthisis, sometimes become affected for several generations with dysentery to the exclusion of phthisis, but by and by, dysentery disappears to give place to phthisis.

4. Hereditary diseases occur to a certain extent independently of external circumstances ; appearing under all sorts of management, and being little affected by changes of locality, separation from diseased stock, or such causes as modify the production of non-hereditary diseases.

5. They are, however, most certainly and speedily developed in circumstances inimical to general good health, and often occur at certain, so called, critical periods of life, when unusual demands on the vital powers take place.

---

* Finlay Dun, V. S., in Journal of the Royal Agricultural Society.

6. They show a striking tendency to modify and absorb into themselves all extraneous diseases; for example, in an animal of consumptive constitution, pneumonia seldom runs its ordinary course, and when arrested, often passes into consumption.

7. Hereditary diseases are less effectually treated by ordinary remedies than other diseases. Thus, although an attack of phthisis, rheumatism or opthalmia may be subdued, and the patient put out of pain and danger, the tendency to the disease will still remain and be greatly aggravated by each attack.

In horses and neat cattle, hereditary diseases do not usually show themselves at birth, and sometimes the tendency remains latent for many years, perhaps through one or two generations and afterwards breaks out with all its former severity."

The diseases which are found to be hereditary in horses are scrofula, rheumatism, rickets, chronic cough, roaring, ophthalmia or inflammation of the eye,—grease or scratches, bone spavin, curb, &c. Indeed, Youatt says, "there is scarcely a malady to which the horse is subject, that is not hereditary. Contracted feet, curb, spavin, roaring, thick wind, blindness, notoriously descend from the sire or dam to the foal."

The diseases which are found hereditary in neat cattle are scrofula, consumption, dysentery, diarrhea, rheumatism and malignant tumors. Neat cattle being less exposed to the exciting causes of disease, and less liable to be overtasked or exposed to violent changes of temperature, or otherwise put in jeopardy, their diseases are not so numerous, and what they have are less violent than in the horse, and generally of a chronic character.

Scrofula is not uncommon among sheep, and it presents itself in various forms. Sometimes it is connected with consumption; sometimes it affects the viscera of the abdomen, and particularly the mesenteric glands in a manner similar to consumption in the lungs. The scrofulous taint has been known to be so strong as to affect the fœtus, and lambs have occasionally been born with it, but much oftener they show it at an early age, and any affected in this way are liable to fall an easy prey to any ordinary or prevailing disease which develops in such with unusual severity. Sheep are also liable to several diseases of the brain and of the respiratory and digestive organs. Epilepsy, or "fits," and rheumatism sometimes occur.

Swine are subject to nearly the same hereditary diseases as sheep. Epilepsy is more common with them than with the latter, and they are more liable to scrofula than any other domestic animals.

When properly and carefully managed, swine are not ordinarily very liable to disease, but when, as too often kept in small, damp, filthy styes, and obliged constantly to inhale noxious effluvia, and to eat unsuitable food, we cannot wonder either that they become victims of disease or transmit to their progeny a weak and sickly organization. Swine are not naturally the dirty beasts which many suppose. "Wallowing in the mire," so proverbial of them, is rather from a wish for protection from insects and for coolness, than from any inherent love of filth, and if well cared for they will be comparatively cleanly.

The practice of close breeding, which is probably carried to greater extent with swine than with any other domestic animal, undoubtedly contributes to their liability to hereditary diseases, and when those possessing any such diseases are coupled, the ruin of the stock is easily and quickly effected, for as already stated, they are propagated by either parent, and always most certainly and in most aggravated form, when occurring in both.

With regard to hereditary diseases, it is eminently true that "an ounce of prevention is worth a pound of cure." As a general and almost invariable rule, animals possessing either defects or a tendency to disease should not be employed for breeding. If, however, for special reasons it seems desirable to breed from one which has some slight defect of symmetry, or a faint tendency to disease, although for the latter it is doubtful

4

if the possession of any good qualities can fully compensate, it should be mated with one which excels in every respect in which the other is deficient, and on no account with one which is near of kin to it.

Notwithstanding the importance due to the subject of hereditary diseases, it is also true that few diseases invariably owe their development to hereditary causes. Even such as are usually hereditary are sometimes produced accidentally, (as of course there must be a beginning to everything,) and in such case, they may, or may not be, transmitted to their progeny. As before shown, it is certain that they sometimes are, which is sufficient reason to avoid such for breeding purposes. It is also well known that, in the horse, for instance, certain forms of limbs predispose to certain diseases, as bone spavin is most commonly seen where there is a disproportion in the size of the limb above and below the hock, and others might be named of similar character; in all such cases the disease may be caused by an agency which would be wholly inadequate in one of more perfect form, but once existing, it is liable to be reproduced in the offspring—all tending to show the great importance of *giving due heed* in selecting breeding animals *to all qualities, both external and internal,* so long as "like produces like."

CHAPTER III.

THE LAW OF VARIATION.

We come now to consider another law, by which that of similarity is greatly modified, to wit, the law of variation or divergence. All organic beings, whether plants or animals, possess a certain flexibility or pliancy of organization, rendering them capable of change to a greater or less extent. When in a state of nature variations are comparatively slow and infrequent, but when in a state of domestication they occur much oftener and to a much greater extent. The greater variability in the latter case is doubtless owing, in some measure, to our domestic productions being reared under conditions of life not so uniform, and different from, those to which the parent species was exposed in a state of nature.

Flexibility of organization in connexion with climate, is seen in a remarkable degree in Indian corn. The small Canada variety, growing only three feet high and ripening in seventy to ninety days when carried southward, gradually enlarges in the whole plant until it may be grown twelve feet high and upwards, and requires one hundred and fifty days to ripen its seed. A southern variety brought northward, gradually dwindles in size and ripens earlier until it reaches a type specially fitted to its latitude.

Variation, although the same in kind, is greater in degree, among domesticated plants than among animals. From the single wild variety of the potato as first discovered and taken to Europe, have sprung innumerable sorts. Kemp, in his work on Agricultural Physiology, tells us, that on the maratime cliffs of England, there exists a little plant with a fusiform root, smooth glaucous leaves, flowers similar to wild mustard and of a saline taste. It is called by botanists, *Brassica oleracea.* By cultivation there have been obtained from this insignificant and apparently useless plant—

1st, all borecoles or kails, 12 varieties or more.

2d, all cabbages having heart.

3d, the various kinds of Savoy cabbages.

4th, Brussels sprouts.

5th, all the broccolis and cauliflowers which do not heart.

6th, the rape plant.

7th, the ruta baga or Swedish turnip.

8th, yellow and white turnips.

9th, hybrid turnips.

10th, kohl rabbi.

Similar examples are numerous among our common useful plants, and among flowers the dahlia and verbena furnish an illustration of countless varieties, embracing numberless hues and combinations of color, from purest white through nearly all the tints of the rainbow to almost black, of divers hights too, and habits of growth, springing up under the hand of cultivation in a few years from plants which at first yielded only a comparatively unattractive and self-colored flower. In brief, it may be said, that nearly or quite all the choicest productions both of our kitchen and flower gardens are due to variations induced by cultivation in a course of years from plants which in their natural condition would scarcely attract a passing glance.

We cannot say what might have been the original type of many of our domestic animals, for the inquiry would carry us beyond any record of history or tradition regarding it, but few doubt that all our varieties of the horse, the ox, the sheep and the dog, sprang each originally from a single type, and that the countless variations are due to causes connected with their domestication. Of those reclaimed within the period of memory may be named the turkey. This was unknown to the inhabitants of the old continent until discovered here in a wild state. Since then, having

4*

been domesticated and widely disseminated, it now offers varieties of wide departure from the original type, and which have been nurtured into self-sustaining breeds, distinguished from each other by the possession of peculiar characteristics.

Among what are usually reckoned the more active causes of variation may be named climate, food and habit.

Animals in cold climates are provided with a thicker covering of hair than in warmer ones. Indeed, it is said that in some of the tropical provinces of South America, there are cattle which have an extremely rare and fine fur in place of the ordinary pile of hair. Various other instances could be cited, if necessary, going to show that a beneficent Creator has implanted in many animals, to a certain extent, a *power of accommodation* to the circumstances and conditions amid which they are reared.

The *supply of food*, whether abundant or scanty, is one of the most active cases of variation known to be within the control of man. For illustration of its effect, let us suppose two pairs of twin calves, as nearly alike as possible, and let a male and a female from each pair be suckled by their mothers until they wean themselves, and be fed always after with plenty of the most nourishing food; and the others to be fed with skimmed milk, hay tea and gruel at first, to be put to grass at two months old, and subsequently fed on coarse and innutritious fodder. Let these be bred from separately, and the same style of treatment kept up, and not many generations would elapse before we had distinct varieties, or breeds, differing materially in size, temperament and time of coming to maturity.

Suppose other similar pairs, and one from each to be placed in the richest blue-grass pastures of Kentucky, or in the fertile valley of the Tees; always supplied with abundance of rich food, these live luxuriously, grow rapidly, increase in hight, bulk, thickness, every way, they early reach the full size which they are capable of attaining; having nothing to induce exertion, they become inactive, lazy, lethargic and fat. Being bred from, the progeny resemble the parents, "only more so." Each generation acquiring more firmly and fixedly the characteristics induced by their situation, these become hereditary, and we by and by have a *breed* exhibiting somewhat of the traits of the Teeswater or Durhams from which the improved Short-horns of the present day have been reared.

The others we will suppose to have been placed on the hill-sides of New England, or on the barren Isle of Jersey, or on the highlands of Scotland, or in the pastures of Devonshire. These being obliged to roam longer for a scantier repast grow more slowly, develop their capabilities in regard to size not only more slowly, but, perhaps, not fully at all—they become more active in temperament and habit, thinner and flatter in muscle. Their young cannot so soon shift for themselves and require more milk, and the dams yield it. Each generation in its turn becomes more completely and fully adapted to the circumstances amid which they are reared, and if bred indiscriminately with any thing and every thing else, we by and by have the common mixed cattle of New England, miscalled natives; or if kept more distinct, we have something approaching the Devon, the Ayrshire, or the Jersey breeds.

A due consideration of the natural effect of climate and food is a point worthy the special attention of the stock-husbandman. If the breeds employed be well adapted to the situation, and the capacity of the soil is such as to feed them fully, profit may be safely calculated upon. Animals are to be looked upon as machines for converting herbage into money. Now it costs a certain amount to keep up the motive power of any machine, and also to make good the wear and tear incident to its working; and in the case of animals it is only so much as is digested and assimilated, *in addition to the amount thus required*, which is converted into meat, milk or wool; so that the greater the proportion which the latter bears to the former, the greater will be the *profit* to be realized from keeping them.

There has been in New England generally a tendency to choose animals of large size, as large as can be had from any where, and if they possess symmetry and all other good qualities commensurate with the size, and if plenty of nutritious food can be supplied, there is an advantage gained by keeping such, for it costs less, other things being equal, to shelter and care for one animal than for two. But our pastures and meadows are not the richest to be found any where, and if we select such as require, in order to give the profit which they are capable of yielding, more or richer food than our farms can supply, or than we have the means to purchase, we must necessarily fail to reap as much profit as we might by the selection of such as could be easily fed upon home resources to the point of highest profit.

Whether the selection be of such as are either larger or smaller than suit our situation, they will, and equally in both cases, vary by degrees towards the fitting size or type for the locality in which they are kept, but there is this noteworthy difference, that if larger ones be brought in, they will not only diminish, but deteriorate, while if smaller be brought in, they will enlarge *and improve*.

The bestowal of food sufficient both in amount and quality to enable animals to develop all the excellencies inherent in them, and to obtain all the profit to be derived from them, is something very distinct from undue forcing or pampering. This process may produce wonderful animals to look at, but neither useful nor profitable ones, and there is danger of thus producing a most undesirable variation, for, as in plants, we find that forcing, pampering, high culture or whatever else it may be called, may be carried so far as to result in the production of double flowers, (an unnatural development,) and these accompanied with greater or less inability to perfect seed, so in animals, the same process may be carried far enough to produce sterility. Instances are not wanting, and particularly among the more recent improved Short-horns, of impotency among the males and of barrenness in the females, and in some cases where they have borne calves they have failed to secrete milk for their nourishment.* Impotency in bulls of various breeds has not unfrequently occurred from too high feeding, and especially if connected *with lack of sufficient exercise.*†

---

* See Rowley's Prize Report on Farming in Derbyshire, in Journal of Royal Agricultural Society, Vol. 14.

† A *working bull*, though perhaps not so pleasing to the eye as a fat one, (for fat sometimes covers a multitude of defects,) is a surer

Habit has a decided influence towards inducing variation. As the blacksmith's right arm becomes more muscular from the habit of exercise induced by his vocation, so we find in domestic animals that use, or the demand created by habit, is met by a development or change in the organization adapted to the requirement. For instance, with cows in a state of nature or where required only to suckle their young, the supply of milk is barely fitted to the requirement. If more is desired, and if the milk be drawn completely and regularly, the yield is increased and continued longer. By keeping up the demand there is induced in the next generation a greater development of the secreting organs, and more milk is given. By continuing the practice, by furnishing the needful conditions of suitable food, &c., and by selecting in each generation those animals showing the greatest tendency towards milk, a breed specially adapted for the dairy may be established. It is just by this mode that the Ayrshires have, in the past eighty or a hundred years, been brought to be what they are, a breed giving more good milk upon a given quantity of food than any other.

It is because the English breeders of modern Shorthorns altogether prefer beef-making to milk-giving prop-

---

stock-getter; and his progeny is more likely to inherit full health and vigor.

erties that they have constantly fostered variation in favor of the one at the expense of the other until the milking quality in many families is nearly bred out. It was not so formerly—thirty years ago the Short-horns (or as they were then usually called, the Dur-hams) were not deficient in dairy qualities, and some families were famous for large yield. By properly directed efforts they might, doubtless, be bred back to milk, but of this there is no probability, at least in Eng-land, for the tendency of modern practice is very strong toward having each breed specially fitted to its use— the dairy breeds for milk and the beef breeds for meat only. The requirements of the English breeder are in some respects quite unlike those of New England farm-ers—for instance, as they employ no oxen for labor there is no inducement to cultivate working qualities even, in connection with beef.

As an illustration of the effect of habit, Darwin* cites the domestic duck, of which he says, "I find that the bones of the wing weigh less, and the bones of the leg more, in proportion to the whole skeleton, than do the same bones in the wild duck ; and I presume that this change may be safely attributed to the domestic duck flying much less and walking more than its wild parent." And again, "not a single domestic animal can be named

---

* In his Origin of Species.

which has not in some country drooping ears, and the view suggested by some authors, that the drooping is due to the disuse of the muscles of the ear, from the animals not being much alarmed by danger, seems probable."

Climate, food and habit are the principal causes of variation which are known to be in any marked degree under the control of man; and the effect of these is, doubtless, in some measure indirect and subservient to other laws, of reproduction, growth and inheritance, of which we have at present very imperfect knowledge. This is shown by the fact that the young of the same litter sometimes differ considerably from each other, though both the young and their parents have apparently been exposed to exactly the same conditions of life; for had the action of these conditions been specific or direct and independent of other laws, if any of the young had varied, the whole would probably have varied in the same manner.

Numberless hypotheses have been started to account for variation. Some hold that it is as much the function of the reproductive system to produce individual differences as it is to make the child like the parents. Darwin says "the reproductive system is eminently susceptible to changes in the conditions of life; and to this system being functionally disturbed in the parents

5

I chiefly attribute the varying or plastic condition of the offspring. The male and female sexual elements seem to be affected before that union takes place which is to form a new being. But why, because the re-productive system is disturbed this or that part should vary more or less, we are profoundly ignorant. Nevertheless we can here and there dimly catch a faint ray of light, and we may feel sure that there must be some cause for each deviation of structure however slight.''

It may be useless for us to speculate here upon the laws which govern variations. The fact that these exist is what the breeder has to deal with, and a most important one it is, for it is this chiefly, which makes hereditary transmission the problem which it is. His aim should ever be *to grasp and render permanent and increase so far as practicable, every variation for the better, and to reject for breeding purposes such as show a downward tendency.*

That this may be done, there is abundant proof in the success which has in many instances attended the well directed efforts of intelligent breeders. A remarkable instance is furnished in the new Mauchamp-Merino sheep of Mons. Graux, which originated in a single animal, a product of the law of variation, and which by skillful breeding and selection has become an established breed of a peculiar type and possessing valuable properties. Samples of the wool of these sheep were shown at the great exhibition in London, in 1851, and attracted much attention. It was also shown at the great recent Agricultural Exhibition at Paris. A correspondent of the *Mark Lane Express*, says:

" One of the most interesting portions of the sheep-show is that of the Mauchamp variety of Merinos, having a new kind of wool, glossy and silky, similar to mohair. This is an instance of an entirely new breed being as it were created from a mere sport of nature. It was originated by Mons. J. L. Graux. In the year 1828, a Merino ewe produced a peculiar ram lamb, having a different shape from the usual Merino, and possessing a long, straight, and silky character of wool. In 1830, M. Graux obtained by this ram one ram and one ewe, having the silky character of wool. In 1831, among the produce were four rams and one ewe with similar fleeces ; and in 1833 there were rams enough of the new sort to serve the whole flock of ewes. In each subsequent year the lambs were of two kinds ; one possessing the curled elastic wool of the old Merinos, only a little longer and finer ; the other like the new breed. At last, the skillful breeder obtained a flock combining the fine silky fleece with a smaller head, broader flanks, and more capacious chest ; and several flocks being crossed with the Mauchamp variety, have produced also the Mauchamp-Merino breed. The pure Mauchamp wool is remarkable for its qualities as a combing-wool, owing to the strength, as well as the length and fineness of the fibre. It is found of great value by the manufacturers of Cashmere shawls and similar goods, being second only to the true Cashmere fleece, in the fine flexible delicacy of the fibre ; and when in combination with Cashmere wool, imparting strength and consistency. The quantity of the wool has now become as great or greater than from ordinary Merinos, while the quality commands for it twenty-five per cent. higher price in the French market. Surely breeders cannot watch too closely any accidental peculiarity of conformation or characteristic in their flocks or herds."

Mons. Vilmorin, the eminent horticulturist of Paris, has likened the law of similarity to the centripetal force, and the law of variation to the centrifugal force ; and in truth their operations seem analogous, and possibly they may be the same in kind, though certainly unlike in this, that they are not reducible to arithmetical calculation and cannot be subjected to definite measurement. His thought is at least a highly suggestive one and may be pursued with profit.

Among the "faint rays" alluded to by Mr. Darwin as throwing light upon the changes dependent on the laws of reproduction, there is one, perhaps the brightest yet seen, which deserves our notice. It is the apparent influence of the male first having fruitful intercourse with a female upon her subsequent offspring by other males. Attention was first directed to this by the following circumstance, related by Sir Everard Home : A young chestnut mare, seven-eighths Arabian, belonging to the Earl of Morton, was covered in 1815 by a Quagga, which is a species of wild ass from Africa, and marked somewhat in the style of a Zebra. The mare was covered but once by the Quagga, and after a pregnancy of eleven months and four days gave birth to a hybrid, which had, as was expected, distinct marks of the Quagga, in the shape of its head, black bars on the legs and shoulders, &c. In 1817, 1818 and 1821, the same mare was covered by a very fine black Arabian horse, and produced successively three foals, and although she had not seen the Quagga since 1816, they all bore his curious and unequivocal markings.

Since the occurrence of this case numerous others of a similar character have been observed, a few of which may be mentioned. Mr. McGillivray says, that in several foals in the royal stud at Hampton Court, got by the horse "Actæon," there were unmistakable marks of the horse "Colonel." The dams of these foals were bred from by Colonel the previous year.

A colt, the property of the Earl of Suffield, got by "Laurel," so resembled another horse, "Camel," that it was whispered and even asserted at Newmarket that he must have been got by "Camel." It was ascertained, however, that the mother of the colt bore a foal the previous year by "Camel."

Alex. Morrison, Esq., of Bognie, had a fine Clydesdale

mare which in 1843 was served by a Spanish ass and produced a mule. She afterwards had a colt by a horse, which bore a very marked likeness to a mule—seen at a distance, every one sets it down at once as a mule. The ears are nine and one-half inches long,—the girth not quite six feet, stands above sixteen hands high. The hoofs are so long and narrow that there is a difficulty in shoeing them, and the tail is thin and scanty. He is a beast of indomitable energy and durability, and highly prized by his owner.

Numerous similar cases are on record,* and it appears to have been known among the Arabs for centuries, that a mare which has first borne a mule, is ever after unfit to breed pure horses ;† and the fact seems now to be perfectly well understood in all the mule-breeding States of the Union.

A pure Aberdeenshire heifer, the property of a farmer in Forgue, was served with a pure Teeswater bull to which she had a first cross calf. The following season the same cow was served with a pure Aberdeenshire bull, the produce was in appearance a cross-bred calf, which at two years old had long horns; the parents were both hornless.

---

*It was long ago stated by Haller, that when a mare had a foal by an ass and afterwards another by a horse, the second offspring begotten by the horse nevertheless approached in character to a mule.

† See Abd el Kader's letter.

A small flock of ewes, belonging to Dr. W. Wells in the island of Grenada, were served by a ram procured for the purpose ;—the ewes were all white and woolly ; the ram was quite different,—of a chocolate color, and hairy like a goat. The progeny were of course crosses but bore a strong resemblance to the male parent. The next season, Dr. Wells obtained a ram of precisely the same breed as the ewes, but the progeny showed distinct marks of resemblance to the former ram, in color and covering. The same thing occurred on neighboring estates under like circumstances.

Six very superior pure-bred black-faced horned ewes, belonging to Mr. H. Shaw of Leochel-Cushnie, were served by a Leicester ram, (white-faced and hornless.) The lambs were crosses. The next year they were served by a ram of exactly the same breed as the ewes themselves. To Mr. Shaw's astonishment the lambs were without an exception hornless and brownish in the face, instead of being black and horned. The third year (1846) they were again served by a superior ram of their own breed, and again the lambs were mongrels, but showed less of the Leicester characteristics than before. Mr. Shaw at last parted from these fine ewes without obtaining a single pure-bred lamb.*

* Journal of Medical Science, 1850.

"It has been noticed that a well bred bitch, if she have been impregnated by a mongrel dog, will not although lined subsequently by a pure dog, bear thorough-bred puppies in the next two or three litters."*

The like occurrence has been noticed in respect of the sow. "A sow of the black and white breed became pregnant by a boar of the wild breed of a deep chestnut color. The pigs produced were duly mixed, the color of the boar being in some very predominant. The sow being afterwards put to a boar of the same breed as herself, some of the produce were still stained or marked with the chestnut color which prevailed in the first litter and the same occurred after a third impregnation, the boar being then of the same kind as herself. What adds to the force of this case is that in the course of many years' observation the breed in question was never known to produce progeny having the slightest tinge of chestnut color.†

The above are a few of the many instances on record tending to show the influence of a first impregnation upon subsequent progeny by other males. Not a few might also be given showing that the same rule holds in the human species, of which a single one will suffice here:—"A young woman residing in Edinburgh, and

---

* Kirke's Physiology.
† Philosophical Transactions for 1821.

born of white parents, but whose mother previous to her marriage bore a mulatto child by a negro man servant, exhibits distinct traces of the negro. Dr. Simpson, whose patient at one time, the young woman was, recollects being struck with the resemblance, and noticed particularly that the hair had the qualities characteristic of the negro."

Dr. Carpenter, in the last edition of his work on physiology, says it is by no means an infrequent occurrence for a widow who has married again to bear children resembling her first husband.

Various explanations have been offered to account for the facts observed, among which the theory of Mr. McGillivray, V. S., which is endorsed by Dr. Harvey, and considered (as we shall presently see) as very probable at least by Dr. Carpenter, seems the most satisfactory. Dr. Harvey says:

" Instances are sufficiently common among the lower animals where the offspring exhibit more or less distinctly over and beyond the characters of the male by which they were begotten, the peculiarities also of a male by which their mother at some former period had been impregnated. * * * Great difficulty has been felt by physiological writers in regard to the proper explanation of this kind of phenomena. They have been ascribed by some to a permanent impression made somehow by the semen of the first male on the genitals and more particularly on the ova of the female:* and by others to an abiding influence exerted by him on the imagination and operating at the time of her connection subsequently with other males and perhaps during her pregnancy; but they seem to be regarded by most physiologists as inexplicable.

Very recently, in a paper published in the Aberdeen Journal, a Veterinary Surgeon, Mr. James McGillivray of Huntley, has offered an explanation which seems to me to be the true one. His theory is that *"when a pure animal of any breed has been pregnant to an animal of a different breed, such pregnant animal is a cross ever after, the purity of her blood being lost in consequence of her connection with the foreign animal, herself* BECOMING A CROSS FOREVER, *incapable of producing a pure calf of any breed."*

Dr. Harvey believes "that while as all allow, a portion of the mother's blood is continually passing by absorption and assimilation into the body of the fœtus, in order to its nutrition and development, a portion of the blood of the fœtus is as constantly passing in like manner into the body of the mother; that as this commingles there with the general mass of the mother's

---

* The late M. A. Cuming, V. S., of New Brunswick, once remarked to the writer, that it might be due to the fact that the nerves of the uterus, which before the first impregnation were in a rudimentary state, were developed under a specific influence from the semen of the first male, and that they might retain so much of a peculiar style of development as to impress upon future progeny by other males the likeness of the first.

own blood, it inoculates her system with the constitutional qualities of the fœtus, and that, as these qualities are in part derived to the fœtus from the male progenitor, the peculiarities of the latter are thereby so ingrafted on the system of the female as to be communicable by her to any offspring she may subsequently have by other males."

In support of this view, Mr. McGillivray cites a case in which there was presented unmistakable evidence that the organization of the placenta admits the return of the venous blood to the mother; and Dr. Harvey, with much force, suggests that the effect produced is analagous to the known fact that constitutional syphilis has been communicated to a female who never had any of the primary symptoms. Regarding the occurrence of such phenomena, Dr. Harvey under a later date says: "since then I have learned that many among the agricultural body in this district are familiar to a degree that is annoying to them with the facts then adduced in illustration of it, finding that after breeding crosses, their cows though served with bulls of their own breed yield crosses still or rather mongrels; that they were already impressed with the idea of contamination of blood as the cause of the phenomenon; that the doctrine so intuitively commended itself to their minds as soon as stated, that they fancied they were told nothing but what they knew before, so just is the observation that truth proposed is much more easily perceived than without such proposal is it discovered."*

Dr. Carpenter, speaking of phenomena analogous to what are here alluded to, says :

" Some of these cases appear referable to the strong impression left by the first male parent upon the female ; but there are others which seem to render it more likely that the blood of the female has imbibed from that of the fœtus, through the placental circulation, some of the attributes which the latter has derived from its male parent, and that the female may communicate these, with those proper to herself, to the subsequent offspring of a different male parentage.   This idea is borne out by a great number of important facts.   *   * As this is a point of great practical importance it may be hoped that those who have the opportunity of bringing observation to bear upon it, will not omit to do so."

In the absence of more general and accurate observations directed to this point, it is impossible to say to what extent the first male produces impression upon subsequent progeny by other males.   There can be no doubt, however, but that such an impression is made. The instances where it is of so marked and obvious a character as in some of those just related may be com-

*Edinburgh Journal Medical Science, 1849.

paratively few, yet there is abundant reason to believe, that although in a majority of cases the effect may be less noticeable, it is not less real, and demands the special attention of all breeders.

Whether this result is to be ascribed to inoculation of the system of the female with the characteristics of the male through the fœtus, or to any other mode of operation, it is obviously of great advantage for every breeder to know it and thereby both avoid error and loss and secure profit. It is a matter which deserves thorough investigation and the observations should be minute and have regard not only to peculiarities of form, but also to qualities and characteristics not so obvious ; for instance there may be greater or less hardiness, endurance or aptitude to fatten. These may be usually more dependent on the dam, but the male is never without a degree of influence upon them, and it is well established that aptitude to fatten is usually communicated by the Short-horn bull to crosses with cattle of mixed or mongrel origin which are often very deficient in this desirable property.

Mr. McGillivray says : "A knowledge of the fact must be of the greatest benefit to the breeder in two ways, positively and negatively. I have known very great disappointment and loss result from allowing an inferior male to serve a first rate female—the useful-

6

ness of such female being thereby forever destroyed. As for the positive benefits arising from the inoculation—they are obvious to any unbiased mind. The black polled and Aberdeenshire cattle common to this country (Scotland) may be, and often are, improved by the following plan: Select a good, well formed, and healthy heifer—put her, in proper season, to a pure Short-horn bull; after the calf to this Durham bull, breed from the cow with bulls of her own breed; occasionally, and most likely the first time, a red calf ultimately having horns will appear even from the polled bull and cow; but in general the calves will be of the same type with the polled parents but with many points improved, and an aptitude to fatten, to come earlier to maturity, &c., such as no one of the pure polled or Aberdeenshire breed ever exhibited in this country, or any other country, however well kept, previous to the introduction of the Short-horn breed. The offspring of these breeds thus improved, when bred from again, will exhibit many points and qualities of excellence similar to the best crosses but retaining much of the hardiness of the original stock, no mean consideration for this changeable and often severe climate. And, moreover, such crosses,—for they are crosses—will command high prices as improved polled or Aberdeenshire cattle. I happen to know of a case where a farmer, from a distance purchased a two year old heifer of the stamp referred to, for the purpose of improving his polled cattle, and for this heifer he paid fifty guineas."

The knowledge of this law* gives us a clue to the cause of many of the disappointments of which practical breeders often complain and to the cause of many variations otherwise unaccountable, and it suggests particular caution as to the first male employed in the coupling of animals, a matter which has often been deemed of little consequence in regard to cattle, inasmuch as fewer heifers' first calves are reared, than of such as are borne subsequently.

Another faint ray of light touching the causes of

---

* A very striking fact may be related in this connection, which while it may or may not have a practical bearing on the breeding of domestic animals, shows forcibly how mysterious are some of the laws of reproduction. It is stated by the celebrated traveler, Count de Strzelecki, in his Physical Description of New South Wales and Van Dieman's Land. "Whenever," he says, "a fruitful intercourse has taken place between an aboriginal woman and an European male, that aboriginal woman is forever after incapable of being impregnated by a male of her own nation, although she may again be fertile with a European." The Count, whose means and powers of observation are of the highest possible order, affirms that " hundreds of instances of this extraordinary fact are on record in the writer's memoranda all recurring invariably under the same circumstances, all tending to prove that the sterility of the female, which is relative only to one and not to the other male is not accidental, but follows laws as cogent though as mysterious as the rest of those connected with generation." The Count's statement is endorsed by Dr. Maun-

variation is afforded us by the fact that the qualities of offspring are not only dependent on the habitual conditions of the parents, but also upon any peculiar condition existing at the time of sexual congress. For instance, the offspring of parents ordinarily healthy and temperate, but begotten in a fit of intoxication, would be likely to suffer permanently, both physically and mentally, from the condition which the parents had temporarily brought upon themselves. On the other hand, offspring begotten of parents in an unusually healthy and active condition of body and mind, would likely be unusually endowed both mentally and physically. The Arabs in breeding horses take advantage of this fact, for before intercourse, both sire and dam are actively exercised, not to weariness, but sufficiently

---

sell of Dublin, Dr. Carmichael of Edinburgh, and the late Prof. Goodsir, who say they have learned from independent sources that as regards Australia, Strzelecki's statement is unquestionable and must be regarded as the expression of a law of nature. The law does not extend to the negro race, the fertility of the negro female not being apparently impaired by previous fruitful intercourse with a European male.

In reply to an inquiry made whether he had ever noticed exceptional cases, the Count says : " It has not come under my cognizance to see or hear of a native female which having a child with a European had afterwards any offspring with a male of her own race."

The Count's statement is suggestive as to the disappearance of the aborigines of some countries. This has often been the subject of severe comment and is generally ascribed to the rum and diseases introduced by the white man. It would now appear that other influences have also been operative.

to induce the most vigorous condition possible. Of this, too, we have proof in the phenomenon sometimes observed by breeders, that a strong mental impression made upon the female by a particular male, will give the offspring a resemblance to him, even though she have no sexual intercourse with him. Of this, Mr. Boswell in his prize essay published in 1828, gives a remarkable instance. He says that Mr. Mustard of Angus, one of the most intelligent breeders he had ever met with, told him that one of his cows chanced to come into season while pasturing on a field bounded by that of one of his neighbors, out of which field an ox jumped and went with the cow until she was brought home to the bull. The ox was white, with black spots, and horned. Mr. Mustard had not a horned beast in his possession, nor one with any white on it. Nevertheless, the produce of the following spring was a black and white calf with horns.

The case of Jacob is often quoted in support of this view, and although many believe some miraculous agency to have been exerted in his case, and though he could say with truth, "God hath taken away the cattle of your father and given them to me," it seems, on the whole, more probable, inasmuch as supernatural agency may never be presumed, except where we know, or have good reason to believe, that natural causes are

6*

insufficient, that God "gave" them, as he now gives to some, riches or honors ; that is to say, by virtue of the operation of natural laws.   If all who keep cattle would exercise a tithe of the patriarch's shrewdness and sagacity in improving their stock, we should see fewer ill-favored kine than at present.

The possibility of some effect being produced by a strong impression at the time of conception, is not to be confounded with the popular error that "marks" upon an infant* are due to a transient, although strong impression upon the imagination of the mother at any period of gestation, which is unsupported by facts and absurd ; but there are facts sufficient upon record to prove that *habitual* mental condition, and especially at an early stage of pregnancy, *may* have the effect to produce some bodily deformity, and should induce great caution.

* Carpenter's Physiology, new edition, page 783.

CHAPTER IV.

ATAVISM, OR ANCESTRAL INFLUENCE.

It may not be easy to say whether this phenomenon is more connected with the law of similarity, or with that of variation. Youatt, in his work on cattle published by the Society for the Diffusion of Useful Knowledge, inclines to the former. He speaks of it as showing the universality of the application of the axiom that "like produces like"—that when this "may not seem to hold good, it is often because the lost resemblance to generations gone by is strongly revived." The phenomenon, or law, as it is sometimes called, of atavism,* or ancestral influence, is one of considerable practical importance, and well deserves careful attention by the breeder of farm stock.

Every one is aware that it is nothing unusual for a child to resemble its grandfather or grandmother or some ancestor still farther back, more than it does either its own father or mother. The fact is too familiar to require the citing of examples. We find the same oc-

---

* From the Latin *Atavus*—meaning any ancestor indefinitely, as a grandmother's great grandfather.

currence among our domestic animals, and oftener in proportion as the breeds are crossed or mixed up. Among our common stock of neat cattle, (*natives*, as they are often called,) originating as they have done from animals brought from England, Scotland, Denmark, France and Spain, each possessing different characteristics of form, color and use; and bred, as our common stock has usually been, indiscriminately together, with no special point in view, no attempt to obtain any particular type or form, or to secure adaptation for any particular purpose, we have very frequent opportunities of witnessing the results of the operation of this law of hereditary transmission. So common indeed is its occurrence, that the remark is often made, that however good a cow may be, there is no telling beforehand what sort of a calf she may have.

The fact is sufficiently obvious that certain peculiarities often lie dormant for a generation or two and then reappear in subsequent progeny. Stockmen often speak of it as "breeding back," or "crying back." The cause of this phenomenon we may not fully understand. A late writer says, "it is to be explained on the supposition that the qualities were transmitted by the grandfather to the father in whom they were *masked* by the presence of some antagonistic or controlling influence, and were thence transmitted to the son in whom the antagonistic influence being withdrawn they manifest themselves. A French writer on Physiology says, if there is not inheritance of paternal characteristics, there is at least an *aptitude* to inherit them, a disposition to reproduce them; and there is always a transmission of this aptitude to some new descendants, among whom these traits will manifest themselves sooner or later.* Mr. Singer, let us say, has a remarkable aptitude for music; but the influence of Mrs. Singer is such that their children inheriting her imperfect ear, manifest no musical talent whatever. These children however have inherited the disposition of the father in spite of its non-manifestation; and if, when they transmit what in them is latent, the influence of their wives is favorable, the grand-children may turn out musically gifted.

The lesson taught by the law of atavism is very plain. It shows the importance of seeking "thorough-bred" or "well-bred" animals; and by these terms are simply meant such as are descended from a line of ancestors in which for many generations the desirable forms, qualities and characteristics have been *uniformly shown.* In such a case, even if ancestral influence does come in

---

* "S'il n'y a pas héritage des caractéres paternels il y a donc au moins *aptitude* à en heriter, disposition à les reproduire, et toujours cette transmission de cette aptitude à des noveau descendants, chez lesquels ces memes caractéres se manifesteront tôt ou tard."—*Longet's* " *Traite de Physiologie,*" ii: 133.

play, no material difference appears in the offspring, the ancestors being all essentially alike. From this stand point we best perceive in what consists the money value of a good "pedigree." It is in the evidence which it brings that the animal is descended from a line all the individuals of which were alike, and excellent of their kind, and so is almost sure to transmit like excellencies to its progeny in turn;—not that every animal with a long pedigree full of high-sounding names is necessarily of great value as a breeder, for in every race or breed, as we have seen while speaking of the law of variation, there will be here and there some which are less perfect and symmetrical of their kind than others; and if such be bred from, they may likely enough transmit undesirable points; and if they be mated with others possessing similar failings, they are almost sure to deteriorate very considerably.

Pedigree is valuable in proportion as it shows an animal to be descended, not only from such as are purely of its own race or breed, but also from such individuals in that breed as were specially noted for the excellencies for which that particular breed is esteemed. Weeds are none the less worthless because they appear among a crop consisting chiefly of valuable plants, nor should deformed or degenerate plants, although they be true to their kind, ever be employed to produce seed.

If we would have good cabbages or turnips, it is needful to select the most perfect and the soundest to grow seed from, and to continue such selection year after year. Precisely the same rule holds with regard to animals.

The pertinacity with which hereditary traits cling to the organization in a latent, masked or undeveloped condition for long after they might be supposed to be wholly "bred out" is sometimes very remarkable. What is known among breeders of Short-horns as the "Galloway alloy," although originating by the employment for only once of a single animal of a different breed, is said to be traceable even now, after many years, in the occasional development of a "smutty nose" in descendants of that family.

Many years ago there were in the Kennebec valley a few polled or hornless cattle. They were not particularly cherished, and gradually diminished in numbers. Mr. Payne Wingate shot the last animal of this breed, (a bull calf or a yearling,) mistaking it in the dark for a bear. During thirty-five years subsequently all the cattle upon his farm had horns, but at the end of that time one of his cows produced a calf which grew up without horns, and Mr. Wingate said it was, in all respects, the exact image of the first bull of the breed brought there.

Probably the most familiar exemplification of clearly

marked ancestral influence among us, is to be found in the ill-begotten, round-breeched calves occasionally, and not very unfrequently, dropped by cows of the common mixed kind, and which, if killed early, make very blue veal, and if allowed to grow up, become exceedingly profitless and unsatisfactory beasts; the heifers being often sterile, the cows poor milkers, the oxen dull, mulish beasts, yielding flesh of very dark color, ill flavor and destitute of fat. They are known by various names in different localities, in Maine as the "Whitten" and "Peter Waldo" breed, in Massachusetts as "Yorkshire" and "Westminster," in New York as the "Pumpkin buttocks," in England as "*Lyery*" or "*Lyery* Dutch," &c., &c.

Those in northern New England are believed to be descended chiefly from a bull brought from Watervliet, near Albany, New York, more than forty years ago, (in 1818,) by the Shakers at Alfred, in York county, Maine, and afterwards transferred to their brethren in Cumberland county. No one who has proved the worthlessness of these cattle can readily believe that any bull of this sort would have been knowingly kept for service since the first one brought into the State, and yet it is by no means a rare occurrence to find calves dropped at the present time bearing unmistakable evidence of that origin.

It seems likely that this disagreeable peculiarity was

first brought into the country by means of some of the early importations of Dutch or of the old Durham breed.

Culley, in speaking of the Short-horns, inclines to the opinion that they were originally from Holland, and himself recollected men who in the early part of their lives imported Dutch cattle into the county of Durham, and of one Mr. Dobinson he says, he was noted for having the best breed of Short-horns of any and sold at high prices. "But afterwards some other persons of less knowledge, going over, brought home some bulls that introduced the disagreeable kind of cattle called *lyery* or *double lyered*, that is, black-fleshed. These will feed to great weight, but though fed ever so long will not have a pound of fat about them, neither within or without, and the flesh (for it does not deserve to be called beef) is as black and coarse grained as horse flesh. No man will buy one of this kind if he knows any thing of the matter, and if he should be once taken in he will remember it well for the future; people conversant with cattle very readily find them out by their round form, particularly their buttocks, which are turned like a black coach horse, and the smallness of the tail; but they are best known to the graziers and dealers in cattle by the *feel* or *touch* of the fingers; indeed it is this nice touch or feel of the hand that in a great measure constitutes the judge of cattle."

7

CHAPTER V.

RELATIVE INFLUENCE OF THE PARENTS.

The relative influence of the male and female parents upon the characteristics of progeny has long been a fertile subject of discussion among breeders. It is found in experience that progeny sometimes resembles one parent more than the other,—sometimes there is an apparent blending of the characteristics of both,—sometimes a noticeable dissimilarity to either, though always more or less resemblance somewhere, and sometimes, the impress of one may be seen upon a portion of the organization of the offspring and that of the other parent upon another portion; yet we are not authorized from such discrepancies to conclude that it is a matter of chance, for all of nature's operations are conducted by fixed laws, whether we be able fully to discover them or not. The same causes always produce the same results. In this case, not less than in others there are, beyond all doubt, fixed laws, and the varying results which we see are easily and sufficiently accounted for by the existence of conditions or modifying influences not fully patent to our observation.

In the year 1825, the Highland Society of Scotland, proposed as the subject of prize essays, the solution of the question, "whether the breed of live stock connected with agriculture be susceptible of the greatest improvement from the qualities conspicuous in the male or from those conspicuous in the female parent?" Four essays received premiums. Mr. Boswell, one of the prize writers, maintained that it is not only the male parent which is capable of most speedily improving the breed of live stock, "but that the male is the parent which we can alone look to for improvement."

His paper is of considerable length and ably written—abounding in argument and illustrations not easily condensed so as to be given here, and it is but justice to add that he also holds that "before the breed of a country can be improved, much more must be looked to than the answer to the question put by the Highland Society—such as crossing, selection of both parents, attention to pedigree, and to the food and care of offspring."

And of crossing, he says, "when I praise the advantage of crossing, I would have it clearly understood that it is only to bring together animals *not nearly related* but always of *the same breed* ; never attempting to breed from a speed horse and a draught mare or vice versa." Crossing of breeds "may do well enough for once, but will end in vexation, if attempted to be prolonged into a line."

Mr. Christian, in his essay, supports the view, that the offspring bears the greatest resemblance to that parent whether male or female, which has exerted the greatest sway of generative influence in the formation of the fœtus, "that any hypothesis which would assign a superiority, or set limits to the influence of either sex in the product of generation is unsound and inadmissible," and he thus concludes—" as therefore it is unsafe to trust to the qualities of any individual animal, male or female, in improving stock, the best bred and most perfect animals of both sexes should be selected and employed in propagation ; there being, in short, no other certain or equally efficacious means of establishing or preserving an eligible breed."

Mr. Dallas, in his essay, starts with the idea that the seminal fluid of the male invests the ovum, the formation of which he ascribes to the female ; and he supports the opinion, that where external appearance is concerned, the influence of the male will be discovered ; but in what relates to internal qualities, the offspring will take most from the female.   He concludes thus :— " When color, quality of fleece, or outward form is wanted, the male may be most depended on for these ; but when milk is the object, when disposition, hardiness, and freedom from diseases of the viscera, and, in short, all internal qualities that may be desired, then the female may be most relied on."

One of the most valuable of these papers was written by the Rev. Henry Berry of Worcestershire, in which, after stating that the question proposed is one full of difficulty and that the discovery of an independent quality such as that alluded to, in either sex, would be attended with beneficial results, he proceeds to show, that it is not to sex, but to high blood, or in other words, to animals long and successfully selected, and bred with a view to particular qualifications, whether in the male or female parent, that the quality is to be ascribed, which the Highland Society has been desirous to assign correctly.

The origin of the prevalent opinion which assigns this power principally to the male, he explains by giving the probable history of the first efforts in improving stock. The greatest attention would naturally be paid to the male, both on account of his more extended services, and the more numerous produce of which he could become the parent; in consequence of which sires would be well-bred before dams. "The ideas entertained respecting the useful qualities of an animal would be very similar and lead to the adoption of a general standard of excellence, towards which it would

7*

be required that each male should approximate; and thus there would exist among what may be termed fashionable sires, a corresponding form and character different from, and superior to, those of the general stock of the country. This form and character would in most instances have been acquired by *perseverance in breeding from animals which possessed the important or fancied requisites*, and might therefore be said to be almost *confirmed* in such individuals. Under these circumstances, striking results would doubtless follow the introduction of these sires to a common stock ; results which would lead superficial observers to remark, that individual sires possessed properties as *males*, which in fact were only assignable to them as *improved* animals."

The opinion entertained by some, that the female possesses the power generally ascribed to the male, he explains also by a reference to the history of breeding: " It is well known to persons conversant with the subject of improved breeding, that of late years numerous sales have taken place of the entire stocks of celebrated breeders of sires, and thus, the females, valuable for such a purpose, have passed into a great number of hands. Such persons have sometimes introduced a cow so acquired to a bull inferior in point of descent and general good qualities, and the offspring is known, in many instances, to have proved superior to the sire by virtue of the dam's excellence, and to have caused a suspicion in the minds of persons not habituated to compare causes with effects, that certain females also possess the property in question."

The writer gives various instances illustrative of his views, in some of which the male only, and in others the female only, was the high-bred animal, in all of which the progeny bore a remarkable resemblance to the well-bred parent.   He says, that where both parents are equally well bred, and of nearly equal individual excellence, it is not probable that their progeny will give general proof of a preponderating power in either parent to impress peculiar characteristics upon the offspring ;—yet in view of all the information we have upon the subject, he recommends a resort to the best males as the most simple and efficacious mode of improving such stocks as require improvement, and the only proceeding by which stock already good can be preserved in excellence.

Mon. Giron* expresses the opinion that the relative age and vigor of the parents exercises very considerable influence, and states as the results of his observation, that the offspring of an old male and a young female resembles the father less than the mother in pro-

* In his work, " De la Generation," Paris, 1828.

portion as the mother is more vigorous and the father more decrepit, and that the reverse occurs with the offspring of an old female and a young male.

Among the more recent theories or hypotheses which have been started regarding the relative influence of the male and female parents, those of Mr. Orton, presented in a paper read before the Farmers' Club at Newcastle upon Tyne, on the Physiology of Breeding, and of Mr. Walker in his work on Intermarriage, as they both arrived to a certain extent, at substantially the same conclusions by independent observations of their own and as these seem to agree most nearly with the majority of observed facts, are deemed worthy of favorable mention.

The conclusions of Mr. Orton, briefly stated,* are, that in the progeny there is no casual or haphazard blending of the parts or qualities of the two parents, but rather that organization is transmitted by halves, or that each parent contributes to the formation of certain structures, and to the development of certain qualities. Advancing a step further, he maintains, that the male parent chiefly determines the external characters, the general appearance, in fact, the outward structure and locomotive powers of the offspring, as the

---

* Quoted, in part, from a paper by Alex. Harvey, M. D., read before the Medical Society of Southampton, June 6th, 1854.

framework, or bones and muscles, more particularly those of the limbs, the organs of sense and skin; while the female parent chiefly determines the internal structures and the general quality, mainly furnishing the vital organs, i. e., the heart, lungs, glands and digestive organs, and giving tone and character to the vital functions of secretion, nutrition and growth. "Not however that the male is without influence on the internal organs and vital functions, or the female without influence on the external organs and locomotive powers of their offspring. The law holds only within certain restrictions, and these form as it were a secondary law, one of limitations, and scarcely less important to be understood than the fundamental law itself."

Mr. Orton relies chiefly on the evidence presented by *hybrids*, the progeny of distinct species, or by crosses between the most distinct varieties embraced within a single species, to establish his law. The examples adduced are chiefly from the former. The *mule* is the progeny of the male ass and the mare; the *hinny*, that of the horse and the she ass. Both hybrids are the produce of the same set of animals. They differ widely, however, in their respective characters—the mule in all that relates to its external characters having the distinctive features of the ass,—the hinny, in the same respects having all the distinctive features of the horse;

while in all that relates to the internal organs and vital qualities, the mule partakes of the character of the horse, and the hinny of those of the ass. Mr. Orton says—"The mule, the produce of the male ass and mare, is essentially a *modified ass :* the ears are those of an ass somewhat shortened ; the mane is that of the ass, erect ; the tail is that of an ass ; the skin and color are those of an ass somewhat modified ; the legs are slender and the hoofs high, narrow and contracted, like those of an ass. In fact, in all these respects it is an ass somewhat modified. The body and barrel, however, of the mule are round and full, in which it differs from the ass and resembles the mare.

The hinny, on the other hand, the produce of the stallion and she ass, is essentially a *modified horse.* The ears are those of a horse somewhat lengthened ; the mane flowing ; the tail bushy, like that of the horse ; the skin is finer, like that of the horse, and the color varies also, like the horse ; the legs are stronger and the hoofs broad and expanded like those of the horse. In fact, in all these respects it is a horse somewhat modified. The body and barrel, however, of the hinny are flat and narrow, in which it differs from the horse and resembles the she ass.

A very curious circumstance pertains to the voice of the mule and the hinny. The mule *brays*, the hinny *neighs.* The why and wherefore of this is a perfect mystery until we come to apply the knowledge afforded us by the law before given. The male gives the locomotive organs, and the muscles are amongst these; the muscles are the organs which modulate the voice of the animal; the mule has the muscular structure of its sire, and brays; the hinny has the muscular structure of its sire, and neighs."

In connexion with these examples Mr. Orton refers to a special feature seen equally in the two instances, and which seems at first sight, a departure from the principle laid down by him. It is this, both hybrids, the mule and the hinny take after the male parents in all their external characters save one, which is *size.* In this respect they both follow the female parents, the mule being in all respects a larger and finer animal than its sire, the ass; the hinny being in all respects a smaller and inferior animal to its sire, the horse, the body and barrel of the mule being large and round, those of the hinny being flat and narrow; both animals being in these particulars the reverse of their respective sires, but both resembling their female parents.

In explanation of this seeming exception is adduced a well known principle in physiology, which is, that the whole bony framework is moulded in adaptation to the softer structures immediately related to it; the muscles covering it in the case of the limbs ; and to the viscera in that of the great cavities which it assists in forming. Accordingly, in perfect accordance with the views above expressed, the *general* size and form which must be mainly that of the *trunk*, will be determined by the size and character of the viscera of the chest and abdomen, and will therefore accord with that of the female parents by whom the viscera in question are chiefly furnished.

The foregoing are the most important of Mr. Orton's statements. He gives, however, numerous additional illustrations from among beasts, birds and fishes, of which we quote only the following:

"The mule and the hinny have been selected and placed first, because they afford the most conclusive evidence and are the most familiar. Equally conclusive, though perhaps less striking instances, may be drawn from other sources. Thus, it has been observed that when the Ancon or Otter sheep were allowed to breed with common ewes, the cross is not a medium between the two breeds, but that the offspring retains in a great measure the short and twisted legs of the sire.

Buffon made a cross between the male goat and the ewe ; the resulting hybrid in all the instances, which were many, were strongly characteristic of the male parent, more particularly in the hair and length of leg. Curious enough, the number of teats in some of the cases corresponded with those of the goat.

A cross between the male wolf and a bitch illustrates the same law; the offspring having a markedly wolfish aspect; skin, color, ears and tail. On the other hand, a cross between the dog and female wolf afforded animals much more dog-like in aspect—slouched ears and even pied in color. If you look at the descriptions and illustrations of these two hybrids, you will perceive at a glance that the doubt arises to the mind in the case of the first, 'what genus of *wolf* is this?' whereas in the case of the second, 'what a curious *mongrel dog!*'

The views of Mr. Walker in his work on Intermarriage, before alluded to, agree substantially with those of Mr. Orton, *so far as regards crossing between different breeds;* but they cover a broader field of observation and in some respects differ. Mr. Walker maintains that when both parents are of the *same breed* that *either parent may transmit either half* of the organization. That when they are of *different varieties* or breeds (and by parity of reasoning the same should hold, strongly, when hybrids are produced by crossing different *species*) and supposing also that both parents are of equal age and vigor, that the *male* gives the *back head and locomotive organs* and the *female* the *face and*

8

nutritive organs—I quote his language: 'when both parents are of the same variety, *one parent communicates the anterior part of the head, the bony part of the face, the forms of the organs of sense* (the external ear, under lip, lower part of the nose and eye brows being often modified) *and the whole of the internal nutritive system,* (the contents of the trunk or the thoracic and abdominal viscera, and consequently the form of the trunk itself in so far as that depends on its contents.)

The resemblance to that parent is consequently found in the forehead and bony parts of the face, as the orbits, cheek bones, jaws, chin and teeth, as well as the shape of the organs of sense and the tone of the voice.

The other parent communicates the posterior part of the head, the cerebel situated within the skull immediately above its junction with the back of the neck, and the whole of the locomotive system; (the bones, ligaments and muscles or fleshy parts.)

The resemblance to that parent is consequently found in the back head, the few more movable parts of the face, as the external ear, under lip, lower part of the nose, eyebrows, and the external forms of the body, in so far as they depend on the muscles as well as the form of the limbs, even to the fingers, toes and nails. * *

It is a fact established by my observations that in animals of the *same variety, either the male or the female parent* may give *either series of organs* as above arranged—that is *either* forehead and organs of sense, together with the vital and nutritive organs, *or* back head, together with the locomotive organs."

To show that among domesticated animals organization is transmitted by halves in the way indicated, and that either parent may give either series of organs, he cites among other instances the account of the Ancon sheep. "When both parents are of the Ancon or Otter breed, their descendants inherit their peculiar appearance and proportions of form. When an Ancon ewe is impregnated by a common ram, the progeny resembles wholly either the ewe or the ram. The progeny of a common ewe impregnated by an Ancon ram follows entirely in shape the one or the other without blending any of the distinguishing and essential peculiarities of both.

'Frequent instances have occurred where common ewes have had twins by Ancon rams; when one exhibited the complete marks and features of the ewe and the other of the ram. The contrast has been rendered singularly striking when one short legged and one long legged lamb produced at a birth have been sucking the dam at the same time.'

As the short and crooked legs or those of opposite form, here indicate the parent giving the locomotive

system, it is evident that one of the twins derived it
from one parent and the other twin from the other
parent;—the parent not giving it, doubtless communi-
cating in each case, the vital or nutritive system."

Where the parents are of different varieties or species,
Mr. Walker says, "The second law, namely, that of
CROSSING, operates where each parent is of a *different
breed*, and where, supposing both to be of equal age
and vigor, the *male* gives the *back head* and *locomotive
organs*, and the *female* the *face* and *nutritive organs*."

After giving numerous illustrations from facts and
many quotations from eminent breeders, he says, "thus,
in crosses of cattle as well as of horses, the male, except
where feebler or of inferior voluntary and locomotive
power, gives the locomotive system, the female the
vital one."

W. C. Spooner, V. S., one of the most eminent au-
thorities of the present day on this subject, and writing
within the past year in the Journal of the Royal Agricultural Society, says:—"The most probable supposition is, that propagation is done by halves, each parent
giving to the offspring the shape of one half of the
body. Thus the back, loins, hind-quarters, general
shape, skin and size follow one parent; and the fore-
quarters, head, vital and nervous system, the other;
and we may go so far as to add, that the former in the

great majority of cases go with the male parent, and the latter with the female. A corroboration of this fact is found in the common system of putting an ordinary mare to a thorough-bred horse; not only does the head of the offspring resemble the dam but the forelegs likewise, and thus it is fortunately the case that the too-frequently faulty and tottering legs of the sire are not reproduced in the foal, whilst the full thighs and hind quarters which belong to the blood-horse are generally given to the offspring. There is however a minority of cases in which the opposite result obtains. That size is governed more by the male parent there is no great difficulty in showing; familiar examples may be found in the pony-mare and the full sized horse, which considerably exceed the dam in size. Again, in the first cross between the small indigenous ewe and the large ram of another improved breed—the offspring is found to approach in size and shape very much to the ram. The mule offspring of the mare also much resembles both in size and appearance its donkey sire. These are familiar examples of the preponderating influence of the male parent, so far as the external form is considered. To show however that size and hight do not invariably follow the male, we need go no further for illustration than the human subject. How often do we find that in the by no means unfrequent case of the union of a tall man with a short woman, the result in some instances is that all the children are tall and in others all short; or sometimes that some are short and others tall. Within our own knowledge in one case, where the father was tall and the mother short, the children, six in number, are all tall. In another instance, the father being short and the mother tall, the children, seven in number, are all of lofty stature. In a third instance, the mother being tall and the father short, the greater portion of the family are short. Such facts as these are sufficient to prove that hight or growth does not exclusively follow either the one parent or the other. Although this is the case, it is also a striking fact that the union of tall and short parents rarely, if ever, produces offspring of a medium size— midway, as it were, between the two parents.

Thus, in the breeding of animals, if the object be to modify certain defects by using a male or female in which such defects may not exist, we cannot produce this desired alteration; or rather it cannot be equally produced in all the offspring, but can only be attained by weeding out those in whom the objectionable points are repeated. We are, however, of opinion that in the majority of instances, the hight in the human subject, and the size and *contour* in animals, is influenced *much more by the male* than the female parent—and on the other hand, that the constitution, the chest and vital organs, and the forehand generally more frequently follow the female."

Dr. Carpenter, the highest authority in Physiology, says "it has long been a prevalent idea that certain parts of the organism of the offspring are derived from the male, and certain other parts from the female parent; and although no universal rule can be laid down upon this point, yet the independent observations which have been made by numerous practical breeders of domestic animals seem to establish that such a *tendency* has a real existence; the characters of the *animal* portion of the fabric being especially (but not exclusively) derived from the male parent, and those of the *organic* apparatus being in like manner derived from the female parent. The former will be chiefly manifested in the external appearance, in the general configuration of the head and limbs, in the organs of the senses (including the skin) and in the locomotive apparatus; whilst the latter show themselves in the size of the body (which is primarily determined by the development of the viscera contained in the trunk) and in the mode in which the vital functions are performed."

On the whole it may be said that the evidence both from observation and the testimony of the best practical breeders goes to show that each parent usually contributes certain portions of the organization to the offspring, and that each has a modifying influence upon the other. Facts also show that the same parent does not always contribute the same portions, but that the order is reversed. Now, as no operation of nature is by accident, but by virtue of *law*, there must be fixed laws here, and there must also be, at times, certain influences at work to modify the action of these laws. Where animals are of distinct species, or of distinct breeds, transmission is usually found to be in accordance with the rule above indicated, i. e. the male gives mostly the outward form and locomotive system, and the female chiefly the interior system, constitution, &c. Where the parents are of the same breed, it appears that the portions contributed by each are governed in large measure by the condition of each in regard to age and vigor, or by virtue of individual potency or superiority of physical endowment.

This *potency* or power of transmission seems to be legitimately connected with high breeding, or the concentration of fixed qualities obtained by continued descent for many generations from such only as possess in the highest degree the qualities desired. On the other hand it must be admitted that there are exceptional cases not easily accounted for upon any theory, and it seems not improbable that in these the modifying influences may be such as to effect what may approximate a reconstruction or new combination of the elements, in a manner analogous to the chemical changes which we know take place in the constituents of vegetables, as for instance, we find that sugar, gum and starch, substances quite unlike in their appearance and uses, are yet formed from the same elements and in nearly or precisely the same proportions, by a chemistry which we have not yet fathomed. Whether this supposition be correct or not, there is little doubt that if we understood fully all the influences at work, and could estimate fairly all the data to judge from, we might predict with confidence what would be the characteristics of the progeny from any given union.

Practically, the knowledge obtained dictates in a most emphatic manner that every stock-grower use his utmost endeavor to obtain the services of the best sires ; that is, *the best for the end and purposes in view*— that he depend chiefly on the sire for outward form and symmetry—that he select dams best calculated to develop the good qualities of the male, depending chiefly upon these for freedom from internal disease, for hardihood, constitution, and generally for all qualities dependent upon the vital or nutritive system.

The neglect which is too common, and especially in breeding horses, to the qualities of the dam, miserably old and inferior females being often employed, cannot be too strongly censured. In rearing valuable horses the dams are not of less consequence than the sires, although their influence upon the progeny be not the same. This is well understood and practiced upon by the Arab, who cultivates endurance and bottom. If his mare be of the true Kochlani breed he will part with her for no consideration whatever, while you can buy his stallion at a comparatively moderate price. The prevalent practice in England and America of cultivating speed in preference to other qualities, has led us to attach greater importance to the male, and the too common neglect of health, vigor, endurance and constitution in the mares has in thousands of cases entailed the loss of qualities not less valuable, and without which speed alone is of comparatively little worth.

CHAPTER VI.

SEX.

With regard to the laws which regulate the sex of progeny very little is known. Many and extensive observations have been made, but without arriving at any definite conclusions. Nature seems to have provided that the number of either sex produced, shall be nearly equal, but by what means this result is attained, has not been discovered. Some physiologists think the sex decided by the influence of the sire, others think it due to the mother. Sir Everard Home believed the *ovum* or germ, previous to impregnation to be of no sex, but so formed as to be equally fitted to become either male or female, and that it is the process of impregnation which marks the sex and forms the generative organs ; that before the fourth month the sex cannot be said to be confirmed, and that it will prove male or female as the tendency to the paternal or maternal type may preponderate.

Mr. T. A. Knight* was of opinion that the sex of progeny depended upon the influence of the female

---

* Philosophical Transactions, 1809.

parent. He says, "The female parent's influence upon the sex of offspring in cows, and I have reason to believe in the females of our other domestic animals, is so strong, that it may, I think, be pronounced nearly positive." He also says, "I have repeatedly proved that by dividing a herd of thirty cows into three equal parts, I could calculate with confidence upon a large majority of females from one part, of males from another, and upon nearly an equal number of males and females from the remainder. I have frequently endeavored to change the habits by changing the male without success." He relates a case as follows:—— "Two cows brought all female offspring, one fourteen in fifteen years, and the other fifteen in sixteen years, though I annually changed the bull. Both however produced one male each, and that in the same year; and I confidently expected, when the one produced a male that the other would, as she did."

M. Giron, after long continued observation and experiment, stated with much confidence, that the general law upon this point was, that the sex of progeny would depend on the greater or less relative vigor of the individuals coupled. In many experiments purposely made, he obtained from ewes more males than females by coupling very strong rams with ewes either too young, or too aged, or badly fed, and more females than males by a reverse choice in the ewes and rams he put together.

Mon. Martegoute, formerly Professor of Rural Economy, in a late communication to the "Journal D'Agriculture Pratique," says that as the result of daily observations at a sheepfold of great importance, that of the Dishley Mauchamp Merinos of M. Viallet at Blanc, he has, if not deceived, obtained some new hints. He states that Giron's law developed itself regularly at the sheepfold in all cases where difference of vigor was observed in the ewes or rams which were coupled; but he adds another fact, which he had observed every year since 1853, when his observations began. This fact consists—

First, In that at the commencement of the rutting season when the ram is in his full vigor he procreated more males than females.

Second, When, some days after, and the ewes coming in heat in great numbers at once, the ram being weakened by a more frequent renewal of the exertion, the procreation of females took the lead.

Third, The period of excessive exertion having passed, and the number of ewes in heat being diminished, the ram also found less weakened, the procreation of males in majority again commenced."

In order to show that the cause of such a result is

9

isolated from all other influences of a nature to be confounded with it, he gives the details of his observations
in a year when the number of births of males and
females were about equal. He also goes on to say,
that, "at the end of each month all the animals at the
sheepfold are weighed separately, and thanks to these
monthly weighings, we have drawn up several tables
from which are seen the diminution or increase in
weight of the different animals classed in various points
of view, whether according to age, sex or the object
for which they were intended.

Two of these tables have been appropriated to bearing ewes—one to those which have borne and nursed
males and the other to those which have borne and
brought up females. The abstract results of these two
tables have furnished two remarkable facts.

First, The ewes that have produced the female lambs
are, on an average, of a weight superior to those that
produced the males; and they evidently lose more in
weight than these last during the suckling period.

Second, The ewes that produce males weigh less, and
do not lose in nursing so much as the others.

If the indications given by these facts come to be
confirmed by experiments sufficiently repeated, two
new laws will be placed by the side of that which Giron
de Bazareingues has determined by his observations

___

and experiments. On the one hand, as, at liberty, or in the savage state, it is a general rule that the predominance in acts of generation belongs to the strongest males to the exclusion of the weak, and as such a predominance is favorable to the procreation of the male sex, it would follow that the number of males would tend to surpass incessantly that of the females, amongst whom no want of energy or power would turn aside from generation, and the species would find in it a fatal obstacle to its reproduction. But, on the other hand, if it was true that the strongest females and the best nurses amongst them produce females rather than males, nature would thus oppose a contrary law, which would establish the equilibrium, and by an admirable harmony would secure the perfection and preservation of the species, by confiding the reproduction of either sex to the most perfect type of each respectively."

CHAPTER VII.

In-and-in Breeding.

It has long been a disputed point whether the system of breeding *in-and-in* or the opposite one of frequent crossing has the greater tendency to maintain or improve the character of stock. The advocates of both systems are earnest and confident of being in the right. The truth probably is, as in some other similar disputes, that both are right and both wrong—to a certain extent, or within certain limits.

The term *in-and-in* is often very loosely used and is variously understood; some, and among these several of the best writers, confine the phrase to the coupling of those of exactly the same blood, i. e. brothers and sisters; while others include in it breeding from parents and offspring, and others still employ the term to embrace those of more distant relationship. For the latter, the term breeding in, or close breeding, is deemed more fitting.

The prevalent opinion is decidedly against the practice of breeding from any near relationships; it being usually found that degeneracy follows, and often to a serious degree; but it is not proved that this degeneracy, although very common and even usual, is yet a necessary consequence. That ill effects follow in a majority of cases is not to be doubted, but this is easily and sufficiently accounted for upon other grounds. In a state of nature animals of near affinities interbreed without injurious results, and it is found by experience that where domesticated animals are of a pure race, or of a distinct, well defined and pure breed, the coupling of those of near affinities is not so often followed by injurious effects as when they are crosses, or of mixed or mongrel origin, like the great majority of the cattle in the country at large. In the latter case breeding in-and-in is *usually* found to result in decided and rapid deterioration. We should consider also that few animals in a state of domestication are wholly free from hereditary defects and diseases, and that these are propagated all the more readily and surely when possessed by both parents, and that those nearly related are more likely than others, to possess similar qualities and tendencies.

If such is to be regarded as the true explanation, it follows that the same method would be also efficacious in perpetuating and confirming good qualities. Such is the fact; and it is well known that nearly all who have achieved eminence as breeders, have availed them-

9*

selves freely of its benefits. Bakewell, the Messrs. Colling, Mr. Mason, Mr. Bates and others, all practiced it. Mr. Bates' rule was, "breed in-and-in from a bad stock and you cause ruin and devastation, they must always be changing to keep even moderately in caste; but *if a good stock* be selected, you may breed in-and-in as much as you please."* Bakewell originated his famous sheep by crossing from the best he could gather from far or near; but when he had obtained such as suited him, he bred exclusively from within his own. As in all breeding from crosses, it was needful to throw out as weeds, a large proportion of the progeny, but by rigidly doing so, and saving none to breed from but such as became more and more firmly possessed of the forms and qualities desired, the weeds gradually became fewer, until at length he fully established the breed; and he continued it, and sustained its high reputation during his life by in-breeding *connected with proper selections for coupling.* After his death, others, not possessing his tact and judgment in making selections, were less fortunate, and in some hands the breed degenerated seriously, insomuch that it was humorously remarked, "there was nothing but a little tallow left." In others it has been maintained by the same method.

* Mr. Bates, although eminent as a breeder, was not infallible in making his selections, and after long continued close breeding, he was compelled to go out of his own herd to procure breeding animals.

Mr. Valentine Barford of Foscote, has the pedigree of his Leicester sheep since the day of Bakewell, in 1783, and since 1810, he has bred entirely from his own flock, sire and dam, without an inter-change of male or female from any other flock. He observes "that his flock being bred from the nearest affinities—commonly called in-and-in breeding—has not experienced any of the ill effects ascribed to the practice." W. C. Spooner, V. S., speaking of Mr. Barford's sheep says, "His flock is remarkably healthy and his rams successful, but his sheep are small."

Mr. Charles Colling, after he procured the famous bull Hubback, selected cows most likely to develop his special excellencies, and from the progeny of these he bred very closely. From that day to this, the Short-horns as a general thing, have been very closely bred,*

---

* Probably few who have not critically examined the facts regarding close breeding in the improved Short-horns are aware of the extent to which it has been carried. On the 28th of March, 1860, at a sale of Short-horns at Milcote, near Stratford upon Avon (England) thirty-one descendants of a cow called "Charmer," bred of Mr. Colling's purest blood, and praised in the advertisement as "capital milkers and very prolific, *not having been pampered*," sold for £2,140, averaging about $350 each, and many of them were calves. The stock was also praised as " offering to the public as much of the pure blood of ' Favorite' as could be found in any herd." With reference to this sale, which also comprised other stock, the Agricultural Gazette, published a few days previous, had some remarks from which the following is extracted:

"It is unquestionable that the ability of a cow or bull to transmit

and the practice has been carried so far, the selections
not always being the most judicious possible, as to re-

the merit either may possess does in a great degree depend upon its
having been inherited by them through a long line of ancestry.
Nothing is more remarkable than the way in which the earlier improvers of the Short-horn breed carried out their belief in this. They
were indeed driven by the comparative fewness of well bred animals
to a repeated use of the same sire on successive generations of his
own begetting, while breeders now-a-days have the advantage of fifty
different strains and families from which to choose the materials of
their herd, but whether it were necessity or choice it is certain that
the pedigree of no pure bred Short-horn can be traced without very
soon reaching many an illustration of the way in which 'breeding
in-and-in' has influenced its character, deepened it, made it permanent, so that it is handed down unimpaired and even strengthened
in the hands of the judicious breeder. What an extraordinary influence has thus been exerted by a single bull on the fortunes of the
Short-horn breed! There is hardly a single choice pure-bred Short-
horn that is not descended from 'Favorite' (252) and not only
descended in a single line—but descended in fifty different lines.
Take any single animal, and this bull shall occur in a dozen of its
preceding generations and repeatedly up to a hundred times! in the
animals of some of the more distant generations. His influence is
thus so paramount in the breed that one fancies he has created it and
that the present character of the whole breed is due the 'accidental'
appearance of an animal of extraordinary endowments on the stage
in the beginning of the present century. And yet this is not so;—he
is himself an illustration of the breeding in-and-in system—his sire
and dam having been half brother and sister, both got by 'Foljambe.'
And this breeding in-and-in has handed down his influence to the
present time in an extraordinary degree. Take for instance, the cow
'Charmer,' from which as will be seen elsewhere, no fewer than
thirty-one descendants are to be sold next Wednesday. She had of
course two immediate parents, four progenitors in the second generation, eight in the third, sixteen in the fourth, the number necessarily
doubling each step farther back. Of the eight bulls named in the
fourth generation from which she was descended, one was by 'Favor-

sult in many cases in delicacy of constitution, and in some where connected with pampering, in sterility.†

Col. Jaques, of the Ten Hills Farm near Boston, imported a pair of Bremen geese in 1822. They were bred together till 1830, when the gander was accidentally killed. Since then the goose bred with her offspring till she was killed by an attack of dogs in 1852. Great numbers were bred during this time, and of course there was much of the closest breeding, yet there was

ite.' She is one-sixteenth 'Favorite' on that account, but the cow to which he was then put was also descended from 'Favorite,' and so are each of the other seven bulls and seven cows which stand on the same level of descent with the gr. gr. g. dam of 'Charmer.' And in fact it will be found on examination that in so far as ' Charmer's' pedigree is known, which it is in some instances to the sixteenth generation, she is not one-sixteenth only but nearly nine-sixteenths of pure Favorite blood. This arises from 'Favorite' having been used repeatedly on cows descended from himself. In the pedigree of ' Charmer' we repeatedly meet with ' Comet'—' Comet' was by ' Favorite' and his dam ' Young Phœnix' was also by 'Favorite ;' with ' George'—' George' was by ' Favorite' and his dam ' Lady Grace' was also by ' Favorite;' with ' Chilton'—' Chilton' was by ' Favorite' and his dam was also by ' Favorite;' with ' Minor'—' Minor' was by ' Favorite' and his dam also was by ' Favorite;' with ' Peeress'—she was by ' Favorite' and her dam also by ' Favorite;' with ' Bright Eyes'—she was by ' Favorite' and her dam also by ' Favorite;' with ' Strawberry'—she was by ' Favorite' and her dam by ' Favorite;' ' Dandy,' ' Moss Rose,' among the cows and ' North Star' among the bulls are also of similar descent.

There is no difficulty therefore in understanding how this name appears repeatedly in any given generation of the pedigree of any given animal of the Short-horn breed.''

† Journal Royal Agricultural Society, volume 20, page 297.

no deterioration, and in fact some of the later ones were larger and better than the first pair.

The same gentleman also obtained a pair of wild geese from Canada in 1818, which with their progeny were bred from without change until destroyed by dogs with the above named in 1852. They continued perfect as at first.

Among gregarious ruminating animals in a state of nature, all who associate in a herd acknowledge a chieftain, or head, who maintains his position by virtue of physical health, strength and general superiority. He not only directs all their movements but is literally the father of the herd. When a stronger than he comes, the post of chieftain and sire is yielded, but in all probability his successor is one of his own sons, who in turn begets offspring by his sisters. The progeny inheriting full health, strength and development, the herd continues in full power and vigor,* and does not degenerate as often happens when man assumes to make the selections, and chooses according to fancy or convenience. The continuance of health, strength and perfect

---

* It may be said with truth, that the average health and vigor of a wild herd is much higher than it would be if the feebler portion of the young were reared, as in a state of domestication, instead of being destroyed by the stronger, or perishing from hardship; but if close breeding be, of itself and necessarily, injurious, the whole herd should gradually fail, which is not found to be the case.

physical development is believed to depend on the *wisdom of the selection, upon the presence of the desirable hereditary qualities, and the absence of injurious ones,* and not upon relationship whether near or remote.

It has fallen within the observation of most persons that in the human race frequent intermarriages in the same family for successive generations often tend to degeneracy of both mind and body; size and vigor diminishing, and constitutional defects and diseases being perpetuated and aggravated; but neither in this case is the result believed to be a necessary and inevitable consequence. Else how could it be, that Infinite Wisdom, whose operations are ever in accordance with the laws of his own institution, in originating a "peculiar people," chosen to be the depositories of intellectual ard physical power, wealth and influence, and who, in spite of oppression without parallel in the world's history, have ever maintained the possession of a goodly share of all these,—would have allowed their first progenitor, Abraham, to marry his near kinswoman Sarah, a half sister, niece or cousin, and Isaac their son to wed his first cousin Rebecca, and Jacob who sprang from that union, to marry first cousins, and their offspring for long generations to intermarry within their own people and tribes alone? At a later period, marriages within certain degrees of consanguinity were forbidden by Divine authority, but not until the peculiar race was fully established, and so far multiplied, as to allow departure from close breeding without change of characteristics, and not improbably the prohibition was even then based more upon moral reasons, or upon man's ignorance or recklessness regarding selection, than upon physical law.

Such laws exist among us at present, and it is well they do, inasmuch as for the reasons already given there is greater probability of degeneracy by means of such connections than among those not so related by blood. But they present an instance of the imperfection of human laws, it being impossible for any legal enactments to prevent wholly the evil thus sought to be avoided. It would be better far, if such a degree of physiological knowledge existed and such caution was exercised among the community generally, as would prevent the contraction of any marriages, where, from the structure and endowments of the parties, debility, deformity, insanity or idiocy must inevitably be the portion of their offspring whether they are more nearly related than through their common ancestor, Noah, or not.

If we adopt Mr. Walker's views, it is easy to see how parents of near affinities may produce offspring perfect and healthy, or the reverse. He holds, that to secure satisfactory results from any union, there should be some inherent, constitutional, or fundamental difference; some such difference as we often see in the human family to be the ground of preference and attachment; as men generally prefer women of a feminine rather than a masculine type. All desire, in a mate, properties and qualities not possessed by themselves. Now assuming as Mr. Walker holds, that organization is transmitted by halves, and that, in animals of the same variety, either parent may give either series of organs, we can see in the case of brother and sister that if one receives the locomotive system of the father and the nutritive system of the mother, and the other the locomotive system of the mother and the nutritive system of the father, they are essentially unlike, there is scarcely any similarity between them, although, as we say, of precisely the same blood; and their progeny if coupled might show no deterioration; whereas, if both have the same series of organs from the same parents, they would be essentially the same, a sort of quasi identity would exist between them, and they are utterly unfit to be mated. There might be impotency, or barrenness, or the progeny, if any, would be decidedly inferior to the parents; and the same applies, more or less, to other relatives descended from a common ancestry, but more distant than brother and sister.

10

Mr. Walker also holds that where the parents are not only of the same variety but of the same family in the narrowest sense, the female always gives the locomotive system and the father the nutritive ; in which case the progeny is necessarily inferior to the parents.

A careful consideration of the subject brings us to the following conclusions, viz :

That in general practice, with the grades and mixed animals common in the country, *close breeding should be scrupulously avoided* as highly detrimental. It is better *always* to avoid breeding from near affinities whenever stock-getters of the same breed and of equal merit can be obtained which are not related. Yet, where this is not possible, or where there is some desirable and clearly defined purpose in view, as the fixing and perpetuating of some valuable quality in a particular animal not common to the breed, and the breeder possesses the knowledge and skill needful to accomplish his purpose, and the animals are perfect in health and development, close breeding may be practiced with advantage.

CHAPTER VIII.

CROSSING.

The practice of crossing, like that of close breeding, has its strong and its weak side. Substantial arguments can be brought both in its favor and against it. Judiciously practiced, it offers a means of procuring animals *for the butcher*, often superior to and more profitable than those of any pure breed. It is also admissible as the foundation of a systematic and well considered attempt to establish a new breed. Such attempts, however, as they necessarily involve considerable expense, and efforts continued during a long term of years, will be rarely made. But when crossing is practiced injudiciously and indiscriminately, and especially when so done for the purpose of procuring *breeding animals*, it cannot be too severely censured, and is scarcely less objectionable than careless in-and-in breeding.

The following remarks, from the pen of W. C. Spooner, V. S., are introduced as sound and reliable, and as comprising nearly all which need be said on the subject of crossing breeds possessing distinctive characteristics :

"Crossing is generally understood to refer to the alliance of animals of different breeds, such as between a thorough-bred and a half-bred among horses or a South Down and Leicester among sheep. Now the advantages or disadvantages of this system depend entirely on the object we have in view, whether merely to beget an animal for the butcher, or for the purpose of perpetuating the species. If the latter is the object, then crossing should be adopted gradually and with care, and by no means between distant or antagonistic qualities, as for example a thorough-bred and a cart-horse. The result of the latter connection is generally an ill-assorted and unfavorable animal, too heavy perhaps for one purpose, and too light for another. If we wish to instil more activity into the cart-horse breed, it is better to do so by means of some half-bred animal, whilst the latter can be improved by means of the three-parts-bred horse and this again by the thorough-bred. There is a remarkable tendency, in breeding, for both good qualities and bad to disappear for one or two generations, and to reappear in the second and third ; thus an animal often resembles the grand dam more than the dam. This peculiarity is itself an objection to the practice of crossing, as it tends to prevent uniformity and to encourage contrarieties ; and thus we find in many flocks and herds that the hopes of the breeders have been entirely baffled and a race of mongrels established.

The first cross is generally successful—a tolerable degree of uniformity is produced, resembling in external conformation the sire, which is usually of a superior breed; and thus the offspring are superior to the dams. These cross-bred animals are now paired amongst each other, and what is the consequence? Uniformity at once disappears; some of the offspring resemble the grandsire, and others the grandams, and some possess the disposition and constitution of the one and some of the other; and consequently a race of mongrels is perpetuated. If, however, the cross is really a good and desirable one, then, by means of rigorous and continued selection, pursued for several generations, that is, by casting aside, as regards breeding purposes, every animal that does not exhibit uniformity, or possess the qualifications we are desirous of perpetuating, a valuable breed of animals may in the course of time be established. By this system many varieties of sheep have been so far improved as to become almost new breeds; as for instance the New Oxford Downs which have frequently gained prizes at the great Agricultural Meetings as being the best long wooled sheep.

To cross, however, merely for crossing sake—to do so without that care and vigilance which we have deemed so essential—is a practice which cannot be too much condemned. It is in fact a national evil and a sin against society, that is, if carried beyond the first cross, or if the cross-bred animals are used for breeding. A useful breed of animals may thus be lost, and a generation of mongrels established in their place, a result

10*

which has followed in numerous instances amongst every breed of animals.

The principal use of crossing, however, is to raise animals for the butcher. In this respect it has not (with sheep) been adopted to the extent which it might to advantage. The male being generally an animal of a superior breed and of a vigorous nature, almost invariably stamps his external form, size and muscular development on the offspring, which thus bear a strong resemblance to him, whilst their internal nature derived from the dam, well adapts them to the locality, as well as to the treatment to which their dams have been accustomed.

With regard to cattle, the system cannot be so advantageously pursued (except for the purpose of improving the size and qualities of the calf, where veal is the object) in as much as every required qualification for breeding purposes can be obtained by using animals of the pure breeds. But with sheep, where the peculiarities of the soil as regards the goodness of feed, and exposure to the severities of the weather, often prevent the introduction of an improved breed, the value of using a new and superior ram is often very considerable, and the weight of mutton is materially increased, without its quality being impaired, while earlier maturity is at the same time obtained. It involves, however, more systematic attention than farmers usually like to bestow, for it is necessary to employ a different ram for each purpose; that is, a native ram for a portion of the ewes to keep up the purity of the breed, and a foreign ram to raise the improved cross-bred animals for fatting either as lambs or sheep. This plan is adopted by many breeders of Leicester sheep, who thus employ South Down rams to improve the quality of the mutton. One inconvenience attending this plan, is the necessity of fatting the maiden ewes as well as the wethers; they may however be disposed of as fat lambs, or the practice of spaying might be adopted, so as to increase the fatting disposition of the animal. Crossing, therefore, should be adopted with the greatest caution and skill where the object is to improve the breed of animals; it should never be practiced carelessly or capriciously, but it may be advantageously pursued with a view to raising superior and profitable animals for the butcher."

In another paper on this subject, after presenting many interesting details regarding British breeds of sheep and the results of crossing, Mr. Spooner says:

"We cannot do better, in concluding our paper, than gather up and arrange in a collected form, the various points of our subject, which appear to be of sufficient importance to be again presented to the attention of our readers. We think, therefore, we are justified in coming to the conclusions:

1st. That there is a direct pecuniary advantage in judicious cross-breeding; that increased size, disposition to fatten, and early maturity, are thereby induced.

2d. That while this may be caused for the most part, by the very fact of crossing, yet it is principally due to the superior influence of the male over the size and external appearance of the offspring; so that it is desirable, for the purpose of the butcher, that the male should be of a larger frame than the female, and should excel in those peculiarities we are desirous of reproducing. Let it be here however, repeated, as an exceptional truth, that though as a rule the male parent influences mostly the size and external form, and the female parent the constitution, general health and vital powers, yet that the opposite result sometimes takes place.

3d. Certain peculiarities may be imparted to a breed by a single cross. Thus, the ponies of the New Forest exhibit characteristics of blood, although it is many years since that a thorough-bred horse was turned into the forest for the purpose. So, likewise, we observe in the Hampshire sheep the Roman nose and large heads, which formed so strong a feature in their maternal ancestors, although successive crosses of the South Down were employed to change the character of the breed. *   *   *

4th. Although in the crossing of sheep for the purpose of the butcher, it is generally advisable to use males of a larger breed, provided they possess a disposition to fatten; yet, in such cases, it is of importance that the *pelvis* of the female should be wide and capacious, so that no injury should arise in lambing, in consequence of the increased size of the heads of the lambs. The shape of the ram's head should be studied for the same reason. In crossing, however, for the purpose of establishing a new breed, the size of the male must give way to other more important considerations; although it will still be desirable to use a large female of the breed which we seek to improve. Thus the South Downs have vastly improved the larger Hampshires, and the Leicester the huge Lincolns and the Cotswolds.

5th. Although the benefits are most evident in the first cross, after which, from pairing the cross-bred animals, the defects of one breed or the other, or the incongruities of both, are perpetually breaking out— yet, unless the characteristics and conformation of the two breeds are altogether averse to each other, nature opposes no barrier to their successful admixture; so that in the course of time, by the aid of selection and careful weeding, it is practicable to establish a new breed altogether. This, in fact, has been the history of our principal breeds. * * *

We confess that we cannot entirely admit either of the antagonistic doctrines held by the rival advocates of crossing and pure breeding. The public have reason to be grateful to the exertions of either party; and still more have they respectively reason to be grateful to each other. * * * *

Let us conclude by repeating the advice that, when equal advantages can be attained by keeping a pure breed of sheep, such pure breed should unquestionably be preferred; and that, although crossing for the purpose of the butcher may be practiced with impunity, and even with advantage, yet no one should do so for the purpose of establishing a new breed, unless he has clear and well defined views of the object he seeks to accomplish, and has duly studied the principles on which it can be carried out, and is determined to bestow for the space of half a life-time his constant and unremitting attention to the discovery and removal of defects."

The term crossing is sometimes used in a much more restricted sense, as in the remark of Mr. Boswell in his essay quoted on page 69 where he says, " When I praise the advantage of crossing I would have it clearly understood that it is only to bring together animals *not nearly related* but always of *the same breed.*" It is evident that such crossing as this is wholly unobjectionable ; no one but an avowed and ultra advocate of close breeding could possibly find any fault with it.

There is yet another style of crossing which when practicable, may, it is believed, be made a means to the highest degree of improvement attainable, and especially in the breeding of horses.   The word "breed" is often used with varying signification.   In order to be understood, let me premise that I use it here simply to designate a class of animals possessing a good degree of uniformity growing out of the fact of a common origin and of their having been reared under similar conditions.   The method proposed is to unite animals *possessing similarity of desirable characteristics, with difference of breed;* that is to say, difference of breed in the sense just specified.   From unions based upon this principle, the selections being guided by a skillful judgment and a discriminating tact, we may expect progeny possessing not only a fitting and symmetrical development of the locomotive system, but also an amount and intensity of nervous energy and power unattainable by any other method.

Such was in all probability the origin of the celebrated horse Justin Morgan; an animal which not only did more to stamp excellence and impart value to the roadsters of New England than any other, but was the originator of the only distinct, indigenous breed of animals of which America can boast;—a breed which as fast and durable road horses and for any light harness work, is not equalled by any other, any where. In the present state of our knowledge it is scarcely conceivable how an animal possessing the endowments of Justin Morgan could have originated in any other way than from such a parentage as above indicated. On the other hand it is very certain that *contrast in character*, as well as in breed, has occasioned much of the disappointment of which breeders have had occasion to complain.

The principle here laid down is one of broad application, and should never be lost sight of in attempts at improvement by crossing. Another point worthy special attention is that all crossing, to insure successful results, should be gentle rather than violent; that is, never couple animals possessing marked dissimilarity, but endeavor to remedy faults and to effect improvement by gradual approaches. Harmony of structure and a proper balancing of desirable characteristics, "an equilibrium of good qualities," as it has been happily expressed, can be secured only in this way.

It may not be out of place here to say, that much of the talk about *blood* in animals, especially horses, is sheer nonsense. When a "blood horse" is spoken of, it means, so far as it means any thing, that his pedigree can be traced to Arabian or Barbary origin, and so is possessed of the peculiar type of structure and great nervous energy which usually attaches to "thorough-bred" horses. When a bull, or cow, or sheep is said to be of "pure blood," it means simply that the animal is of some distinct variety—that it has been bred from an ancestry all of which were marked by the same peculiarities and characteristics.

So long as the term "blood" is used to convey the idea of definite hereditary qualities it may not be objectionable. We frequently use expressions which are not strictly accurate, as when we speak of the sun's rising and setting, and so long as every body knows that we refer to apparent position and not to any motion of the sun, no false ideas are conveyed. But to suppose that the hereditary qualities of an animal attach to the blood more than to any other fluid or to any of the tissues of the body, or that the blood of a high-bred horse is essentially different from that of another, is entirely erroneous. The qualities of an animal depend upon its organization and endowments, and the blood is only the vehicle by which these are nourished and sustained;—moreover the blood varies in quality, composition and amount, according to the food eaten, the air breathed and the exercise taken. If one horse is better than another it is not because the fluid in his veins is of superior quality, but rather because his structure is more perfect mechanically, and because nervous energy is present in fitting amount and intensity.

For illustration, take two horses—one so built and endowed that he can draw two tons or more, three miles in an hour; the other so that he can trot a mile in three minutes or less. Let us suppose the blood coursing in the veins of each to be transferred to the other; would the draft horse acquire speed thereby, or the trotter acquire power? Just as much and no more as if you fed each for a month with the hay, oats and water intended for the other.

It is well to attend to pedigree, for thus only can we know what are the hereditary qualities, but it is not well to lay too much stress upon "blood." What matters it that my horse was sired by such a one or such a one, if he be himself defective? In breeding horses, *structure* is first, and endowment with nervous energy is next to be seen to, and then pedigree—afterwards that these be fittingly united, by proper selection for coupling, in order to secure the highest degree of probability which the nature of the case admits, that the offspring may prove a perfect machine and be suitably endowed with motive power.

"The body of an animal is a piece of mechanism, the moving power of which is the vital principle, which like fire to the steam engine sets the whole in motion; but whatever quantity of fire or vital energy may be applied, neither the animal machine nor the engine will work with regularity and effect, unless the individual parts of which the machine is composed are properly adjusted and fitted for the purposes for which they are intended; or if it is found that the machine does move by the increase of moving power, still the motion is irregular and imperfect; the bolts and joints are continually giving way, there is a continued straining of the various parts, and the machine becomes worn out and useless in half the time it might have lasted if the proportions had been just and accurate. Such is the case with the animal machine. It is not enough that it is put in motion by the noblest spirit or that it is nourished by the highest blood; every bone must have its just proportion; every muscle or tendon its proper pulley; every lever its proper length and fulcrum; every joint its most accurate adjustment and proper lubrication; all must have their relative proportions and strength, before the motions of the machine can be accurate, vigorous and durable. In every machine modifications are required according as the purposes vary to which it is applied. The heavy dray horse is far from having the arrangement necessary for the purposes of the turf, while the thorough-bred is as ill adapted for the dray. Animals are therefore to be selected for the individual purposes for which they are intended, with the modifications of form proper for the different uses to which they are to be applied; but for whatever purpose they may be intended, there are some points which are common to all, in the adjustment of the individual parts. If the bones want their due proportions, or are imperfectly placed—if the muscles or tendons want their proper levers—if the flexions of the joints be interrupted by the defectiveness of their mechanism, the animal must either be defective in motion or strength; the bones have irregular pressure, and if they do not break, become diseased; if the muscles or tendons do not become sprained or ruptured, they are defective in their action ; if friction or inflammation does not take place in the joints, the motions are awkward and grotesque. As in every other machine, the beauty of the animate, whether in motion or at rest, depends upon the arrangement of the individual parts."

CHAPTER IX.

BREEDING IN THE LINE.

The preferable style of breeding for the great majority of farmers to adopt, is neither to cross, nor to breed from close affinities, (except in rare instances and for some specific and clearly understood purpose,) but to *breed in the line*, that is, select the breed or race best adapted to fulfill the requirements demanded, whether it be for the dairy, for labor or for beef in cattle, or for such combination of these as can be had without too great sacrifice of the principal requisite; whether for fine wool as a primary object and for meat as a secondary one, or for mutton as a primary and wool for a secondary object, and then procure a *pure bred* male of the kind determined on, and breed him to the females of the herd or of the flock; and if these be not such as are calculated to develop his qualities, endeavor by purchase or exchange to procure such as will. Let the progeny of these be bred to another *pure bred* male of the same breed, but as distantly related to the first as may be. Let this plan be steadily pursued, and although we cannot, without the intervention of well bred females, obtain stock purely of kind desired, yet in several generations, if proper care be given in the selection of males, that each one be such as to retain and improve upon the points gained by his predecessor, the stock for most practical purposes will be as good as if thorough-bred.   Were this plan generally adopted, and a system of letting or exchange of males established, the cost might be brought within the means of most persons, and the advantages which would accrue would be almost beyond belief.

The writer on Cattle in the Library of Useful Knowledge well remarks :—" At the outset of his career, the farmer should have a clear and determined conception of the object that he wishes to accomplish.   He should consider the nature of his farm ; the quality, abundance or deficiency of his pasturage, the character of the soil, the seasons of the year when he will have plenty or deficiency of food, the locality of his farm, the market to which he has access and the produce which can be disposed of with greatest profit, and these things will at once point to him the breed he should be solicitous to obtain.   The man of wealth and patriotism may have more extensive views, and nobly look to the general improvement of cattle ; but the farmer, with his limited means and with the claims that press upon him, regards his cattle as a valuable portion of his own little property, and on which every thing should appear to be in natural keeping, and be turned to the best advantage. The best beast for him is that which suits his farm the best, and with a view to this, he studies, or ought to study, the points and qualities of his own cattle, and those of others. The dairyman will regard the quantity of milk—the quality—its value for the production of butter and cheese—the time that the cow continues in milk—the character of the breed for quietness, or as being good nurses—the predisposition to garget or other disease, or dropping after calving—the natural tendency to turn every thing to nutriment—the ease with which she is fattened when given up as a milker, and the proportion of food requisite to keep her in full milk or to fatten her when dry. The grazier will consider the kind of beast which his land will bear—the kind of meat most in demand in his neighborhood—the early maturity—the quickness of fattening at any age— the quality of the meat—the parts on which the flesh and fat are principally laid—and more than all the hardihood and the adaptation to the climate and soil.

In order to obtain these valuable properties the good farmer will make himself perfectly master of the characters and qualities of his own stock. He will trace the connection of certain good qualities and certain bad ones, with an almost invariable peculiarity of shape and structure; and at length he will arrive at a clear conception, not so much of beauty of form (although that is a pleasing object to contemplate) as of that outline and proportion of parts with which *utility* is oftenest combined. Then carefully viewing his stock he will consider where they approach to, and how far they wander from, this utility of form; and he will be anxious to preserve or to increase the one and to supply the deficiency of the other. He will endeavor to select from his own stock those animals that excel in the most valuable points, and particularly those which possess the greatest number of these points, and he will unhesitatingly condemn every beast that manifests deficiency in any one important point. He will not, however, too long confine himself to his own stock, unless it be a very numerous one. The breeding from close affinities has many advantages to a certain extent. It was the source whence sprung the cattle and sheep of Bakewell and the superior cattle of Colling; and to it must also be traced the speedy degeneracy, the absolute disappearance of the New Leicester cattle, and, in the hands of many agriculturists, the impairment of constitution and decreased value of the New Leicester sheep and of the Short-horns. He will therefore seek some change in his stock every second or third year, and that change is most conveniently effected by introducing a new bull. This bull should be of the same breed, and pure, coming from a similar pasturage and climate, but possessing no relationship—or, at most, a very distant one—to the stock to which he is introduced. He should bring with him every good point which the breeder has labored to produce in his stock, and if possible, some improvement, and especially in the points where the old stock may have been somewhat deficient, and most certainly he should have no manifest defect of form; and that most essential of all qualifications, a hardy constitution, should not be wanting.

There is one circumstance, however, which the breeder occasionally forgets, but which is of as much importance to the permanent value of his stock as any careful selection of animals can be—and that is, good keeping. It has been well said that 'all good stock must be both bred with attention and well fed. It is necessary that these two essentials in this species of improvement should always accompany each other; for without good resources of keeping, it would be vain to attempt supporting a valuable stock.' This is true with regard to the original stock. It is yet more evident when animals are absurdly brought from a better to a poorer soil. The original stock will deteriorate if neglected and half-starved, and the improved breed will lose ground even more rapidly, and to a far greater extent."

A very brief resumé of the preceding remarks may be expressed as follows :

The Law of Similarity teaches us to select animals for breeding which possess the desired forms and qualities in the greatest perfection and best combination.

Regard should be had not only to the more obvious characteristics, but also to such hereditary traits and tendencies as may be hidden from cursory observation and demand careful and thorough investigation.

From the hereditary nature of all characteristics, whether good or bad, we learn the importance of having all desirable qualities and properties *thoroughly inbred*; or, in other words, so firmly fixed in each generation, that the next is warrantably certain to present nothing worse,—that no ill results follow from breeding back towards some inferior ancestor,—that all undesirable traits or points be, so far as possible, *bred out*.

So important is this consideration, that in practice, it is decidedly preferable to employ a male of ordinary external appearance, provided his ancestry be all which is desired, rather than a grade or cross-bred animal, although the latter be greatly his superior in personal beauty.

A knowledge of the Law of Divergence teaches us to avoid, for breeding purposes, such animals as exhibit variations unfavorable to the purpose in view ; and to endeavor to perpetuate every real improvement gained ;

also to secure as far as practicable, the conditions necessary to induce or to perpetuate any improvement, such as general treatment, food, climate, habit, &c.

Where the parents do not possess the perfection desired, selections for coupling should be made with critical reference to correcting the faults or deficiencies of one by corresponding excellence in the other.

But to correct defects too much must not be attempted at once. Pairing those very unlike, oftener results in loss than in gain. Mating a horse for speed with a draft mare, will more likely beget progeny good for neither, than for both. Avoid all extremes, and endeavor by moderate degrees to obtain the object desired.

Crossing, between different breeds, for the purpose of obtaining animals for the shambles, may be advantageously practiced to considerable extent, but not for the production of breeding animals. As a general rule cross-bred males should not be employed for propagation, and cross-bred females should be served by thorough-bred males.

In ordinary practice, breeding from near relationships is to be *scrupulously avoided;* for certain purposes, under certain conditions and circumstances, and in the hands of a skillful breeder, it may be practiced with advantage, but not otherwise.

In a large majority of cases (other things being equal) we may expect in progeny the outward form and general structure of the sire, together with the internal qualities, constitution and nutritive system of the dam; each, however, modified by the other.

Particular care should always be taken that the male by which the dam first becomes pregnant is the best which can be obtained; also, that at the time of sexual congress both are in vigorous health.

Breeding animals should not be allowed to become fat, but always kept in thrifty condition; and such as are intended for the butcher should never be fat but once.

In deciding with what breeds to stock a farm, endeavor to select those best adapted to its surface, climate, and degree of fertility; also with reference to probable demand and proximity to markets.

No expense incurred in procuring choice animals for propagation, or any amount of skill in breeding, can supersede, or compensate for, a lack of liberal feeding and good treatment. The better the stock, the better care they deserve.

CHAPTER X.

CHARACTERISTICS OF VARIOUS BREEDS.

The inquiry is frequently made, what is the best breed of cattle, sheep, &c., for general use. In reply it may be said that no breed can by any possibility fulfill all requirements in the best possible manner; one is better for meat and early maturity, another for milk, another for wool, and so on. Because under certain circumstances it may be necessary or advisable for a man to serve as his own builder, tailor, tanner and blacksmith, it by no means follows that all which is required will be as well, or as easily done, as by a division of labor. So it is better for many reasons, and more profit can be made, by employing different breeds for different purposes, than by using one for all, and towards such profitable employment we should constantly aim. At the same time there is a large class of farmers so situated that they cannot keep distinct breeds, and yet wish to employ them for different uses, and whose requirements will best be met by a kind of cattle, which, without possessing remarkable excellence in any one direction, shall be sufficiently hardy,

12

the oxen proving docile and efficient laborers for a while, and then turn quickly into good beef upon such food as their farms will produce, the cows giving a fair quantity and quality of milk for the needs of the family and perhaps to furnish a little butter and cheese for market.

Before proceeding to answer the inquiry more definitely, it may be well to remark further, that among the facts of experience regarding cattle, sheep and horses, nothing is better established than that no breed can be transferred from the place where it originated, and to which it was suited, to another of unlike surface, climate and fertility, and retain equal adaptation to its new situation, nor can it continue to be what it was before. It must and will vary. The influence of climate alone, aside from food and other agencies in causing variation, is so great that the utmost skill in breeding, and care in all other respects, cannot wholly control its modifying effects.

It is also pretty well established that no breed brought in from abroad can be fully as good, *other things being equal*, as one indigenous to the locality, or what approximates the same thing, as one, which by being reared through repeated generations on the spot has become thoroughly acclimated; so that the presumption is strongly in favor of *natives*.

When we look about us however, we find, if we except the Morgan horses, nothing which deserves the name of indigenous breeds or races. The cattle and sheep known as "natives" are of mixed foreign origin, and have been bred with no care in selection, but crossed in every possible way. They possess no fixed hereditary traits, and although among them are many of very respectable qualities, and which possess desirable characteristics, they cannot be relied upon *as breeders, to produce progeny of like excellence.* Instead of constancy, there is continual variation, and frequent "breeding back," exhibiting the undesirable traits of inferior ancestors. That a breed might be established from them, by careful selection continued during repeated generations, aided perhaps by judicious crossing with more recent importations, fully as good as any now existing, is not to be doubted. Very probably, a breed for dairy purposes might be thus created which should excel any now existing in Europe, for some of our so called native cows, carelessly as they have been bred, are not surpassed by any of foreign origin upon which great care has been expended. To accomplish this is an object worthy the ambition of those who possess the skill, enthusiasm, ample means and indomitable perseverance requisite to success. But except the single attempt of Col. Jaques, of the Ten Hills

Farm, to establish the Creampot breed,* of which, as little has been heard since his death, it is fair to presume that it has dropped into the level of common grade cattle, no systematic and continued effort has come to our knowledge. Consequently such as may be deemed absolutely the best is a thing of the future; they do not yet exist—and there is no probability that the desideratum will soon be attained. We Yankees are an impatient people; we dislike to wait, for any thing, or to invest where five, ten, twenty or fifty years may be expected to elapse before satisfactory dividends may be safely anticipated.

Still, if all would begin to-day, to use what skill and judgment they have, or can acquire, in breeding only from the best of such as they have, coupling with reference to their peculiarities, and consigning to the butcher as fast as possible every inferior animal, and if, in addition, they would do what is equally necessary, namely, improve their general treatment as much as lies in their power, there would result an immediate, a marked and a steadily progressive improvement in stock. To the acclimation or Americanization already acquired, would be added increased symmetry of form and greater value in many other respects. This is within the power of

---

* This was commenced by a cross of Cœlebs, a Short-horn bull, upon a common cow of remarkable excellence.

every man, and whatever else he may be obliged to leave undone, for want of ability, none should be content to fall short of this. Those who have the command of ample means will of course desire that improvement should be as rapid as possible. They will endeavor at once to procure well bred animals, or in other words, such as already possess the desired qualities so thoroughly inwrought into their organization that they can rely with a good degree of confidence on their imparting them to their progeny.

It may be well to allude here to a distinction between breeds and races. By *breeds*, are understood such varieties as were originally produced by a cross or mixture, like the Leicester sheep for example, and subsequently established by selecting for breeding purposes only the best specimens and rejecting all others. In process of time deviations become less frequent and greater uniformity is secured; but there remains a tendency, greater or less in proportion to the time which elapses and the skill employed in selection, to resolve itself into its original elements, to breed back toward one or other of the kinds of which it was at first composed.

By *races*, are understood such varieties as were moulded to their peculiar type by natural causes, with no interference of man, no intermixture of other varieties, and have continued substantially the same for a

12*

period beyond which the memory and knowledge of man does not reach. Such are the North Devon cattle, and it is fortunate that attention was drawn to the merits of this variety before facilities for inter-communication had so greatly increased as of late, and while yet the race in some districts remained pure. All that breeders have done to better it, is by selections and rejections from within itself; and so, much improvement has been effected without any adulteration. Consequently we may anticipate that so long as no crossing takes place, there will be little variation.

Among the established breeds of cattle the IMPROVED SHORT-HORNS are the most fashionable, and the most widely diffused; and where the fertility of the soil, and the climate, are such as to allow the development of their peculiar excellencies, they occupy the highest rank as a meat-producing breed. Their beef is hardly equal in quality to that of the Devons, Herefords or Scots, the fat and lean being not so well mixed together and the flesh of coarser grain. But they possess a remarkable tendency to lay on fat and flesh, attaining greater size and weight, and coming earlier to maturity than any other breed. These properties, together with their symmetry and stately beauty, make them very popular in those counties of England, where they originated, and wherever else they have been carried, provided their surroundings are such as to meet their wants. In the rich pastures of Kentucky and in some other parts of the west, they seem as much at home as on the banks of the Tees, and are highly and deservedly esteemed. The Short-horns have also been widely and successfully used to cross with most other breeds, and with inferior mixed cattle, as they are found to impress strongly upon them their own characteristics.

Without entering into the question of its original composition, or of its antiquity, regarding both of which much doubt exists, it may suffice here to say, that about a hundred years ago, Charles Colling and others entered zealously and successfully into an attempt to improve them by careful breeding, in whose hands they soon acquired a wide spread fame and brought enormous prices; and the sums realized for choice specimens of this breed from that time to the present, have been greater than for those of any other. Much of their early notoriety was due to the exhibition of an ox reared by Charles Colling from a common cow by his famous bull "Favorite," and known as the "Durham" ox, and also as the "Ketton" ox, (both which names have since then been more or less applied to the breed, but which are now mostly superseded by the original and more appropriate one of Short-horn,) which was shown in most parts of England and Scotland from 1801 to 1807, and whose live weight was nearly four thousand pounds, and which was at one time valued for purposes of exhibition as high as $10,000.

The old Teeswater cattle were remarkably deep milkers, and although it does not appear that good grazing points necessarily conflict with excellence for the dairy, the fact is, that as improvement in feeding qualities was gained, the production of milk in most cases fell off; and although some families at the present time embrace many excellent milkers, the majority of them have deteriorated in this respect about in proportion to the improvement effected as meat-producing animals. The earlier Short-horns introduced into this country were from the very best milking families, and their descendants have usually proved valuable for dairy purposes—but many of those more recently imported are unlike them in this respect. By crossing the males upon the common cows of the country the progeny inherited increased size and symmetry of form, more quiet dispositions, greater aptitude to feed and earlier maturity. Notwithstanding the prejudices with which they were at first received, they gradually rose in estimation, more of them have been introduced than of any other breed, and probably more of the improvement which has taken place in cattle for the last forty years is due to them than to any other; yet *as a pure breed they are not adapted to New England wants.* Their size is beyond the ability of most farms to support profitably: crossed upon such as through neglect in breeding, scanty fare and exposure were bad feeders, too small in size, and too slow in growth, they effected great improvement in all these respects; and this improvement demanded and encouraged the bestowal of more food and better treatment, and so they prospered;—inheriting their constitutions chiefly from the hardy and acclimated dams, the grades were by no means so delicate and sensitive as the pure bred animals to the cold and changes of a climate very unlike that of the mild and fertile region where they originated.

The lethargic temperament characteristic of the Short-horn and which in the grades results in the greater quietness and docility so highly valued, necessarily unfits them for active work; pure bred animals being altogether too sluggish for profitable labor. This temperament is inseparably connected with their aptitude to fatten and early maturity, and these both demand abundant and nutritious food beyond the ability of many to supply and at the same time are incompatible with the activity of habit and hard service demanded of the working ox.

The NORTH DEVONS are deemed to be of longer stand-
ing than any other of the distinct breeds of England,
and they have been esteemed for their good qualities
for several centuries. Mr. George Turner, a noted
breeder of Devons, describes them as follows :—" Their
color is generally a bright red, but varying a little either
darker or more yellow; they have seldom any white
except about the udder of the cow or belly of the bull,
and this is but little seen.   They have long yellowish
horns, beautifully and gracefully curved, noses or muz-
zles white, with expanded nostrils, eyes full and prominent, but calm, ears of moderate size and yellowish
inside, necks rather long, with but little dewlap, and
the head well set on, shoulders oblique with small
points or marrow bones, legs small and straight and
feet in proportion.   The chest is of moderate width,
and the ribs round and well expanded, except in some
instances, where too great attention has been paid to
the hind quarters at the expense of the fore, and which
has caused a falling off, or flatness, behind the shoulders.
The loins are first rate, wide, long and full of flesh, hips
round and of moderate width; rumps level and well
filled at the bed; tail full near the rump and tapering
much at the top.   The thighs of the cows are occasionally light, but the bull and ox are full of muscle, with a
deep and rich flank.   On the whole there is scarcely

any breed of cattle so rich and mellow in its touch, so silky and fine in its hair, and altogether so handsome in its appearance, as the North Devon, added to which they have a greater proportion of weight in the most valuable joints and less in the coarse, than any other breed, and also consume less food in its production.

As milkers they are about the same as most other breeds;—the general average of a dairy of cows being about one pound of butter per day from each cow during the summer months, although in some instances the very best bred cows give a great deal more.

As working oxen they greatly surpass any other breed. They are perfectly docile and excellent walkers, are generally worked until five or six years old, and then fattened at less expense than most other oxen."

The author of the report on the live stock shown at the exhibition of the Royal Agricultural Society at Warwick in 1859 (Mr. Robert Smith) says :

"Although little has been written on it, the improvement of the Devon has not been neglected ; on the contrary, its breeding has been studied like a science, and carried into execution with the most sedulous attention and dexterity for upwards of two hundred years. The object of the Devon breeder has been to lessen those parts of the animal frame which are least

useful to man, such as the bone and offal, and at the
same time to increase such other parts (flesh and fat) as
furnish man with food.   These ends have been accomplished by a judicious selection of individual animals
possessing the wished for form and qualities in the
highest degree, which being perpetuated in their progeny in various proportions, and the selection being
continued from the most approved specimens among
these, enabled the late Mr. Francis Quartly at length to
fully establish the breed with the desired properties.
This result is substantially confirmed by the statistics
contained in Davy's 'Devon Herd-Book.'   We have
been curious enough to examine these pedigrees, and
find that nine-tenths of the present herds of these truly
beautiful animals are directly descended (especially in
their early parentage) from the old Quartly stock.
Later improvements have been engrafted on these by
the Messrs. Quartly of the present day.   The example
of various opulent breeders and farmers in all parts of
the country has tended to spread this improvement, by
which the North Devon cattle have become more general and fashionable.   The leading characteristics of
the North Devon breed are such as qualify them for
every hardship.   They are cast in a peculiar mold, with
a degree of elegance in their movement which is not to
be excelled.   Their hardihood, resulting from compact-

ness of frame and lightness of offal, enables them (when wanted) to perform the operations of the farm with a lively step and great endurance. For the production of animal food they are not to be surpassed, and in conjunction with the Highland Scot of similar pretension, they are the first to receive the attention of the London West-end butcher. In the show-yard, again, the form of the Devon and its rich quality of flesh serve as the leading guide to all decisions. He has a prominent eye, with a placid face, small nose and elegantly turned horns, which have an upward tendency (and cast outward at the end) as if to put the last finish upon his symmetrical form and carriage. These animals are beautifully covered with silken coats of a medium red color. The shoulder points, sides, and foreflanks are well covered with rich meat, which, when blended with their peculiar property of producing meat of first-rate quality along their tops, makes them what they are—'models of perfection.' Of course, we here speak of the best-bred animals. Some object to the North Devon, and class him as a small animal, with the remark, 'He is too small for the grazier.' In saying this it should ever be remembered that the Devon has its particular mission to perform, viz., that of converting the produce of cold and hilly pastures into meat, which could not

13

be done to advantage by large-framed animals, however good their parentage."

The Devons have been less extensively, and more recently, introduced than the Short-horn, but the experience of those who have fairly tried them fully sustains the opinions given above, and they promise to become a favorite and prevailing breed. The usual objection made to them by those who have been accustomed to consider improvement in cattle to be necessarily connected with enlargement of size, is, that they are too small. But their size instead of being a valid objection, is believed to be a recommendation, the Devons being as large as the fertility of New England soils generally are *capable of feeding fully and profitably.*

Their qualities as working oxen are unrivalled, no other breed so uniformly furnishing such active, docile, strong and hardy workers as the Devons, and their uniformity is such as to render it very easy to match them. Without possessing so early maturity as the Short-horns, they fatten readily and easily at from four to six years old, and from their compact build and well balanced proportions usually weigh more than one accustomed to common cattle would anticipate.

The Devons are not generally deep milkers but the milk is richer than that of most other breeds, and some families, where proper care and attention have been given to this quality in breeding, yield largely. It is, however, as a breed for general use, combining beef, labor and milk, in fair proportion, that the Devons will generally give best satisfaction, as they are hardy enough to suit the climate, and cheaply furnish efficient labor and valuable meat.

Farmers, whose ideas upon stock have been formed wholly from their experience with Short-horns and their grades, have often been surprised at witnessing the facility with which Devons sustain themselves upon scanty pasturage, and not a few when first critically examining well bred specimens, sympathize with the feeling which prompted the remark made to the reporter of the great English Exhibition at Chester, after examining with him fine specimens of the Devons—"I am delighted; I find we Short-horn men have yet much to learn of the true formation of animals; their beautiful contour and extreme quality of flesh surprise me."

The HEREFORDS are an ancient and well established breed, and are probably entitled to be called a race. Little is known with certainty of their origin beyond the fact that for many generations they can be traced as the peculiar breed of the county whence they derive their name. Youatt says that "Mr. Culley, although an excellent judge of cattle, formed a very erroneous opinion of the Herefords when he pronounced them to be nothing but a mixture of the Welsh with a bastard race of Long Horns. They are evidently an aboriginal breed, and descended from the same stock as the Devon. If it were not for the white face and somewhat larger head and thicker neck it would not at all times be easy to distinguish between a heavy Devon and a light Hereford."

Mr. Gisborne says "The Hereford brings good evidence that he is the British representative of a widely diffused and ancient race. The most uniform drove of oxen which we ever saw, consisted of five hundred from the Ukraine. They had white faces, upward horns and tawny bodies. Placed in Hereford, Leicester or Northampton markets, they would have puzzled the graziers as to the land of their nativity; but no one would have hesitated to pronounce that they were rough Herefords."

Mr. Rowlandson, in his prize report on the farming of Herefordshire, says "The Herefords, or as they have sometimes been termed, the middle horned cattle have ever been esteemed a most valuable breed, and when housed from the inclemency of the weather, probably put on more meat and fat in proportion to the food consumed, than any other variety. They are not sc hardy as the North Devon cattle, to which they bear a general resemblance; they however are larger than the Devons, especially the males. On the other hand, the Herefords are larger boned, to compensate for which defect, may be cast in the opposite scale the fact that the flesh of the Hereford ox surpasses all other breeds for that beautiful marbled appearance caused by the intermixture of fat and lean which is so much prized by the epicure. The Hereford is usually deeper in the chine, and the shoulders are larger and coarser than the Devon. They are worse milkers than the Devon, or than, perhaps, any other breed, for the Hereford grazier has neglected the female and paid the whole of his attention to the male." It is said that formerly they were of a brown or reddish brown color, and some had grey or mottled faces. Mr. P. Tully states that the white face originated accidentally on a farm belonging to one of his ancestors. "That about the middle of the last century the cow-man came to the house announcing as a remarkable fact that the favorite cow had produced a white faced bull calf. This had never been known to have occurred before, and, as a curiosity it was agreed that the animal should be kept and reared as a future sire. Such, in a few words, is the origin of a fact that has since prevailed through the country, for the progeny of this

13*

very bull became celebrated for white faces." Of late years there has been much uniformity of color; the face, throat, the under portion of the body, the inside and lower part of the legs and the tip of the tail being white, and the other parts of the body a rich deep red.

Compared with the Short-horn the Hereford is nearly as large, of rather less early maturity, but a better animal for grazing, and hardier. The competition between these breeds in England is very close and warm, and taking many facts together it would seem probable that the Hereford is in some instances rather more profitable, and the Short-horn generally more fashionable. Challenges have been repeatedly offered by Hereford men to Shorthorn men to feed an equal number of each in order to test their respective merits, and have usually been declined, perhaps because if the decision was against them, the loss might be serious, and if they won, the gain would be little or nothing, the Short-horns being more popular already and commanding higher prices.

As working oxen the Herefords are preferable to the Short-horns, being more hardy and active. Some complaint is made of their being "breachy." Their large frames demand food, and if enough be furnished they are content, but if not, they have intelligence and activity enough to help themselves if food be within reach. Their chief merit is as large oxen, for heavy labor, and for beef. Some grade cows from good milking dams give a fair quantity of milk, and what they give is always rich, but wherever they have been introduced, milking qualities generally deteriorate very much.

The AYRSHIRES are a breed especially valuable for dairy purposes. Regarding its origin, Mr. Aiton who felt much interest in the subject, and whose opportunities for knowing the facts were second to those of no other, writing about forty years since, says, "The dairy breed of cows in the county of Ayr now so much and so deservedly esteemed, is not, in their present form, an ancient or indigenous race, but a breed formed during the memory of living individuals and which have been gradually improving for more than fifty years past, till now they are brought to a degree of perfection that has never been surpassed as dairy stock in any part of Britain, or probably in the world. They have increased to double their former size, and they yield about four and some of them five times as much milk as formerly. By greater attention to breeding and feeding, they have been changed from an ill-shaped, puny, mongrel race of cattle to a fixed and specific breed of excellent color and quality. So gradually and imperceptibly were improvements in the breed and condition of the cattle introduced, that although I lived in Ayrshire from 1760 to 1785, and have traversed it every year since, I have difficulty in stating from my own observation or what I have learned from others, either the precise period when improvement began, or the exact means by which a change so important was wrought." He then relates several instances in which between 1760 and 1770 some larger cows were brought in of the English or Dutch breeds, and of their effect he says, "I am disposed to believe that although they rendered the red color with white patches fashionable in Ayr, they could not have had much effect in changing the breed into their present highly improved condition," and thinks it mainly due to careful selections and better treatment.

Mr. Aiton says "the chief qualities of a dairy cow are that she gives a copious draught of milk, that she fattens readily and turns out well in the shambles. In all these respects combined the Ayrshire breed excels all others in Scotland, and is probably superior to any in Britain. They certainly yield more milk than any other breed in Europe. No other breed fatten faster, and none cut up better in the shambles, and the fat is as well mixed with the lean flesh, or marbled, as the butchers say, as any other. They always turn out better than the most skillful grazier or butcher who are strangers to the breed could expect on handling

them. They are tame, quiet, and feed at ease without roaming, breaking over fences, or goring each other. They are very hardy and active, and are not injured but rather improved by lying out all night during summer and autumn."

Since Mr. Aiton wrote, even greater care and attention has been paid to this breed than before, and it is now well entitled to rank as the first dairy breed in the world, quantity and quality of yield and the amount of food required being all considered. Compared with the Jersey, its only rival as a dairy breed, the milk of the Ayrshire is much more abundant, and richer in caseine, but not so rich in oily matter, although better in this respect than the average of cows.

Experience of their qualities in this country shows that if they do not here fully sustain their reputation in Scotland, they come near to it, as near as the difference in our drier climate allows, giving more good milk upon a given amount of food than any other. Upon ordinarily fertile pastures they yield largely and prove very hardy and docile. The oxen too are good workers, fatten well, and yield juicy, fine flavored meat.

The JERSEY race, formerly known as the Aldernay, is almost exclusively employed for dairy purposes, and may not be expected to give satisfaction for other uses. Their milk is richer than that of any other cows, and the butter made from it possesses a superior flavor and a deep rich color, and consequently commands an extraordinary price in all markets where good butter is appreciated.

The Jersey cattle are of Norman origin, and until within about twenty or thirty years were far more un-inviting in appearance than now, great improvement having been effected in their symmetry and general appearance by means of careful selections in breeding, and this without loss of milking properties. The cows are generally very docile and gentle, but the males when past two or three years of age often become vicious and unmanageable. It is said that the cows fatten readily when dry, and make good beef.

There is no branch of cattle husbandry which prom-ises better returns than the breeding and rearing of milch cows. Here and there are to be found some good enough. In the vicinity of large towns and cities are many which having been culled from many miles around, on account of dairy properties, are considera-bly above the average, but taking the cows of the country together they do not compare favorably with the oxen. Farmers generally take more pride in their oxen, and strive to have as good or better than any of their neighbors, while if a cow will give milk enough to rear a large steer calf and a little besides, it is often deemed satisfactory.

SHEEP.—The sheep first introduced into this country were of English origin, and generally not very dissimilar to the ancient unimproved Down sheep. Probably some were these—as many of the first cattle were the Devons of that day. More than fifty years since the Merinos were introduced and extensively bred. At various periods other choice breeds have been introduced. The number kept has fluctuated very much, depending mainly on the market value of wool. When it was high many kept sheep, and when it fell the flocks were neglected.

The true mission of the sheep in fulfilling the threefold purpose of furnishing *food, and raiment, and the means of fertilization,* seems not yet to be generally apprehended. One of the most serious defects in the husbandry of New England at the present time, is the prevalent neglect of sheep. Ten times the present number might be easily fed, and they would give in meat, wool and progeny, more direct profit than any other domestic animal, and at the same time the food they consume would do more towards fertilizing the farms than an equal amount consumed by any other animal.

It is notorious that our pastures have seriously deteriorated in fertility and become overrun with worthless weeds and bushes to the exclusion of nutritious grasses.

Sheep husbandry has declined. If these two facts as uniformly stand to each other in the relation of cause and effect, as they certainly do in many instances, the remedy is suggested at once—replace the animal with "golden feet." After devoting the best land to cultivation and the poorest to wood, we have thousands upon thousands of acres evidently intended by the Creator for sheep walks, because better adapted for this purpose than for any other. An indication of Providence so unmistakable as this should not be unheeded.

The MERINOS are perhaps the most ancient race of sheep extant. They originated in Spain, and were for ages bred there alone. In 1765 they were introduced into Saxony, where they were bred with care and with special reference to increasing the fineness of the wool, little regard being paid to other considerations. They were also taken to France and to Silesia, and from all these sources importations have been made into the United States. The Spanish Merino has proved the most successful, and by skill and care in breeding has been greatly improved, insomuch that intelligent judges are of opinion that some of the Vermont flocks are superior to the best in Europe, both in form, hardiness, quantity of fleece and staple. They are too well known to require a detailed description here. Suffice it to say that they are below rather than above medium size, possessing a good constitution, and are thrifty, and cheaply kept. Their chief merit is as fine wooled sheep, and as such they excel all others. As mutton sheep they are constitutionally and anatomically deficient, being of late maturity and great longevity, (a recommendation as fine wooled sheep,) having too flat sides, too narrow chests, too little meat in the best parts, and too great a percentage of offal when slaughtered. Their mutton, however, is of fair quality when mature and well fatted. As nurses they are inferior to many other breeds. Many careful, extensive and protracted attempts have been made to produce a breed combining the fleece of the Merino with the carcass of the Leicester or other long wooled sheep. They have all signally failed. The forms, characteristics and qualities of breeds so unlike seem to be incompatible with one another. A cross of the Merino buck and Leicester ewe gives progeny which is of more rapid growth than the Merino alone, and is hardier than the Leicester. It is a good cross for the butchers' use, but not to be perpetuated. Improvement in the Merino should be sought by skillful selection and pairing the parents in view of their relative fitness to one another.

The LEICESTER, or more properly the New Leicester, is the breed which Bakewell established, and is repeatedly referred to in the preceding pages. It has quite superseded the old breed of this name. His aim was to produce sheep which would give the greatest amount of meat in the shortest time on a given amount of food, and for early maturity and disposition to fatten, it still ranks among the highest. The objections to the breed for New England are, that they are not hardy enough for the climate, and require richer pastures and more abundant food than most farmers can supply. Its chief value in such locations is for crossing upon ordinary sheep for lambs and mutton.

The COTSWOLDS derive their name from a low range of hills in Gloucestershire. These have long been noted for the numbers and excellence of the sheep there maintained, and are so called from Cote, a sheepfold, and Would, a naked hill. An old writer says:— " In these woulds they feed in great numbers flocks of sheep, long necked and square of bulk and bone, by reason (as is commonly thought) of the weally and hilly situation of their pastures, whose wool, being most fine and soft, is held in passing great account amongst all nations." Since his time, however, great changes have passed both upon the sheep and the district they inhabit. The improved Cotswolds are among the largest British breeds, long wooled, prolific, good nurses, and of early maturity. More robust, and less liable to disease than the Leicesters, of fine symmetry and carrying great weight and light offal, they are among the most popular of large mutton sheep.

The SOUTH DOWN is an ancient British breed, taking its name from a chalky range of hills in Sussex and other counties in England about sixty miles in length, known as the South Downs, by the side of which is a tract of land of ordinary fertility and well calculated for sheep walks, and on which probably more than a million of this breed of sheep are pastured. The flock tended by the "Shepherd of Salisbury Plain," of whose earnest piety and simple faith Hannah More has told us in her widely circulated tract, were South Downs. Formerly these sheep possessed few of the attractions they now present. About the year 1782 Mr. John Ellman of Glynde turned his attention to their improvement. Unlike his cotemporary Bakewell, he did not attempt to make a new breed by crossing, but by attention to the principles of breeding, by skillful selections for coupling and continued perseverance for fifty years, he obtained what he sought—health, soundness of constitution, symmetry of form, early maturity, and facility of fattening, and thus brought his flock to a high state of perfection. Before he began we are told that the South Downs were of "small size and ill shape, long and thin in the neck, high on the shoulders, low behind, high on the loins, down on the rumps, the tail set on very low, sharp on the back, the ribs flat," &c., &c., and were not mature enough to fatten until three years old or past.　Of his flock in 1794, Arthur Young* says: " Mr. Ellman's flock of sheep, I must observe in this place, is unquestionably the first in the country; there is nothing that can be compared with it; the wool is the finest and the carcass the best proportioned; although I saw several noble flocks afterwards which I examined with a great degree of attention; some few had very fine wool, which might be equal to his, but then the carcass was ill-shaped, and many had a good carcass with coarse wool; but this incomparable farmer had eminently united both these circumstances in his flock at Glynde.　I affirm this with the greater degree of certainty, since the eye of prejudice has been at work in this country to disparage and call in question the quality of his flock, merely because he has raised the merit of it by unremitted attention above the rest of the neighboring farmers, and it now stands unrivalled." This, it will be noticed, was only twelve years after he began his improvements.　To Mr. Ellman's credit be it said that he exhibited none of the selfishness which characterized Mr. Bakewell's career, but was always

---

* Annals of Agriculture, Vol. 11, p. 224.

ready to impart information to those desirous to learn, and labored zealously to encourage general improvement. That he was pecuniarily successful is evident from the continued rise in the price of his sheep. The Duke of Richmond, Mr. Jonas Webb, Mr. Grantham, and other cotemporaries and successors of Mr. Ellman have carried successfully forward the work so well begun by him. The Improved South Downs now rank first among British breeds in hardiness, constitution, early maturity, symmetry, and quality of mutton and of wool combined. The meat usually brings one to two cents per pound more than that of most other breeds in Smithfield market. It is of fine flavor, juicy, and well marbled. The South Downs are of medium size, (although Mr. Webb has in some cases attained a live weight in breeding rams of 250 pounds, and a dressed weight of 200 pounds in fattened wethers,) hardy, prolific, and easily kept, suceeding on short pastures, although they pay well for liberal feeding.

The OXFORD DOWNS may be named as an instance of successful cross-breeding. They originated in a cross between the Improved Cotswolds and the Hampshire Downs.* Having been perpetuated now for more than

---

* The Hampshires are somewhat larger than the South Downs, and quite as hardy—the fleece a trifle shorter. The Oxford Downs are not to be confounded with the New Oxfordshires.

twenty years, they possess so good a degree of uniformity as to be entitled to the designation of a distinct breed, and have lately been formally recognized as such in England. They were first introduced into Massachusetts by R. S. Fay, Esq., of Lynn, and into Maine by Mr. Sears, both in 1854. They were first bred with a view to unite increased size with the superiority of flesh and patience of short keep which characterize the Downs. It is understood that they inherit from the Cotswold a carcass exceeding in weight that of the Downs from a fifth to a quarter; a fleece somewhat coarser but heavier than that of the Downs by one-third to one-half; and from the latter they inherit rotundity of form and fullness of muscle in the more valuable parts, together with the brown face and leg.

In reply to a note of inquiry addressed to Mr. Fay, he says: "I selected the Oxford Downs with some hesitation as between them and the Shropshire Downs, after a careful examination of all the various breeds of sheep in England. My attention was called to them by observing that they took, (1854,) without any distinct name, all the prizes as mutton sheep at Birmingham and elsewhere, where they were admitted to compete. They were only known under the name of half or cross bred sheep, with name of the breeder. Mr. Rives of Virginia and myself went into Oxfordshire to look at them, and so little were they known as a class, that Philip Pusey, Esq., President of the Royal Agricultural Society, knew nothing about them, although one of his largest tenants, Mr. Druce, had long bred them. It is only within two years that they were formally recognized at a meeting, I believe, of the Smithfield club, and they then received the name which I gave them years ago, of Oxford Downs. By this name they are now known in England. I can only add that an experience of six years confirms all that is claimed for them. Fifty-two ewes produced seventy-three healthy lambs from February 13th to March 15th, this year. The same ewes sheared an average of more than seven pounds to the fleece, unwashed wool, which sold for 34 cents per pound. A good ram should weigh as a shearling from 180 to 250 pounds; a good ewe from 125 to 160 pounds. They fatten rapidly, and thrive on rough pasture. My flock, now the older and poorer ones have been disposed of, will average, I have no doubt, eight pounds of wool to the fleece. The mutton is exceedingly fine and can be turned into cash in 18 months from birth."

The kind of sheep most desirable, on the whole, in any given case, depends chiefly on the surface, character and fertility of the farm and its location. At too great a distance from a good meat market to allow of a profitable sale of the carcass, the Spanish Merino is doubtless to be preferred, but if nearer, the English breeds will pay better. Mutton can be grown cheaper than any other meat. It is daily becoming better appreciated, and strange as it may seem, good mutton brings a higher price in our best markets than the same quality does in England. Its substitution in a large measure for pork would contribute materially to the health of the community.

Winter fattening of sheep may often be made very profitable and deserves greater attention, especially where manure is an object—(and where is it not?) In England it is considered good policy to fatten sheep if the increase of weight will pay for the oil cake or grain consumed; the manure being deemed a fair equivalent for the other food, that is, as much straw and turnips as they will eat. Lean sheep there usually command as high a price per pound in the fall as fatted ones in the spring, while here the latter usually bear a much higher price, which gives the feeder a great advantage. The difference may be best illustrated by a simple calculation. Suppose a wether of a good mutton breed weighing 80 pounds in the fall to cost 6 cents per pound ($4.80) and to require 20 pounds of hay per week, or its equivalent in other food, and to gain a pound and a half each week, the gain in weight in four months would be about 25 pounds, which at 6 cents per pound would be $1.50 or less than $10 per ton for the hay consumed; but if the same sheep could be bought in fall for 3 cents per pound and sold in spring for 6 cents, the gain would amount to $3.90 or upwards of $20 per ton for the hay—the manure being the same in either case.

For fattening it is well to purchase animals as large and thrifty and in as good condition as can be done at fair prices ; and to feed liberally so as to secure the most rapid increase which can be had without waste of food.

The fattening of sheep by the aid of oil cake or grain purchased for the purpose, may often be made a cheaper and altogether preferable mode of obtaining manure than by the purchase of artificial fertilizers, as guano, superphosphate of lime, &c. It is practiced extensively and advantageously abroad and deserves at least a fair trial here.

HORSES.—It does not seem necessary in this connection to give descriptions of the various breeds of horses, as comparatively few of our animals can fairly be said to be of any pure or distinct varieties. Names are common enough, but the great majority of the horses among us are so mixed in their descent from the breeds which have been introduced at various times from abroad, as to be almost as near of kin to one as to another. Success in breeding will depend far more upon attention to selection in regard to structure and endowments than to names. Although it may be somewhat beyond the scope of an attempt to treat merely of the principles of breeding to offer remarks regarding its practice, a few brief hints may be pardoned; and first, let far more care be taken in respect of breeding mares. Let none be bred from which are too old, or of feeble constitution, or the subjects of hereditary disease. No greater mistake can be made than to suppose that a mare fit for nothing else, is worthy to be bred from. If fit for this, she is good for much else—gentle, courageous, of good action, durable and good looking; outward form is perhaps of less importance than in the male, but serious defect in this greatly lessens her value. She should be *roomy*, that is the pelvis should be such that she can well develop and easily carry and deliver the foal. Youatt says, "it may, perhaps, be justly affirmed that there is more difficulty in selecting a good mare to breed from, than a good horse, because she should possess somewhat opposite qualities. Her carcass should be long to give room for the growth of the fœtus, yet with this there should be compactness of form and shortness* of leg.''

The next point is the selection of a stallion. It is easy enough to say that he should be compactly built, "having as much goodness and strength as possible condensed in a little space," and rather smaller relatively than the mare, that he should be of approved descent and possess the forms, properties and characteristics which are desired to be perpetuated. It is not very difficult to specify with tolerable accuracy what forms are best adapted for certain purposes, as an oblique shoulder, and depth, rather than width, of chest are indispensable for trotting; that in a draft horse this obliquity of shoulder is not wanted, one more upright being preferable, and so forth; but after all, a main point to secure success is *relative adaptation of the parents to each other,* and here written directions are necessarily insufficient and cannot supply the place of skill and judgment to be obtained only by careful study and practical experience; nor is it always easy, even if

---

* Mr. Youatt here probably refers to length below, rather than above, the knee and hock.

fully aware of the necessary requirements, to find them in the best combination in the horses nearest at hand. A stallion may be all which can be desired for one dam and yet be very unsuitable for another. In this aspect we can perceive how valuable results may accrue from such establishments as now exist in various sections of the country, where not a single stallion only is kept, but many, and where no pains nor expense are spared to secure the presence of superior specimens of the most approved breeds, and choice strains of blood in various combinations; so that the necessary requirements in a sire are no sooner fairly apprehended than they are fully met. On this point therefore, my suggestion is, that this relative adaptation of the parents to one another be made the subject of patient and careful study; and a word of caution is offered lest in the decisions made, too great importance be attached to speed alone. That speed is an element of value is not doubted, nor do I intimate that he who breeds horses to sell, may not aim to adapt his wares to his market as much as the man who breeds neat cattle and sheep, or the man who manufactures furniture to sell. But I do say that speed may be, and often has been, sought at too dear a rate, and that bottom, courage, docility and action are equally elements of money value and equally worthy of being sought for in progeny. Nor is it unlikely that an attempt to breed for these last named qualities, with a proper reference to speed, would result in the production of as many fast horses as we now get, and in addition to this, a much higher average degree of merit in the whole number reared.

Another suggestion may not be out of place. Hitherto (if we except fast trotting) there has been little attention paid to breeding for special purposes, as for draft horses, carriage horses, saddle horses, etc., and the majority of people at the present time undoubtedly prefer horses of all work. This is well enough so long as it is a fact that the wants of the masses are thus best met, but it is equally true that as population increases in density and as division of labor is carried farther, it will be good policy to allow the horse to share in this division of labor, and to breed with reference to different uses; just as it is good policy for one man to prepare himself for one department of business and another for another. The same principle holds in either case.

Sufficient attention has never been paid to the breaking and training of horses. Not one in a thousand receives a proper education. It ought to be such as to bring him under perfect control, with his powers fully developed, his virtues strengthened and his vices eradicated. What usually passes for breaking is but a distant approximation to this. The methods recently promulgated by Rarey and Baucher are now attracting attention, and deservedly too, not merely for the immediate profit resulting from increased value in the subjects, but in view of the ultimate results which may be anticipated ; for, as we have seen when treating of the law of similarity, acquired habits may in time become so inbred as to be transmissible by hereditary descent.